KHNUM-PTAH
TO
COMPUTER

THE AFRICAN INITIALIZATION OF COMPUTER SCIENCE

By

AFRICAN CREATION ENERGY

DECEMBER 12, 2012

WWW.AFRICANCREATIONENERGY.COM

ISBN 978-1-300-49891-9

Printed in the United States of America

In Loving Memory of
J e s s e   S t o n e w a l l
(November 1915 – December 2012),
my Grandfather and first teacher of Electronics,
thank you for all that you have passed on to me

Hut-Ka-Ptah

At Dusk and Dawn,
During the Twilight Hours,
Sometimes the Sun and the Moon
Shine at the Same Time

Dedicated to the Future Generation of African Computer Programmers, African Software Designers, African Robotics Engineers, African Mechanics Masons, and African Virtual Reality Architects. Know that Computer Science, and the many fields that are derived from it, is a Craft that is one of many operative expressions of your traditional African culture. Let this book provide you with the mental link to your Ancient and Traditional African customs and cultural past as you create and design the highly-advanced technologies for the survival, well-being, and improved quality of life for African people in the future.

African Creation Energy

Creative Solution-Based Technical Consulting

TABLE OF CONTENTS

0000 – INITIALIZATION

The fundamentals of Computer Science started in Africa!

If considered with a myopic mentality that has been socialized and programmed by individuals with complete disdain for everything African, then the preceding statement may seem quite audacious. However, when traditional African culture, philosophy, cosmology, customs, sciences, and technologies are thoroughly examined, we find a plethora of empirical evidence throughout the African continent over time which substantiates the reality that the fundamentals of Computer Science did indeed originate in Africa. In this book, what is meant by the term **"fundamentals of Computer Science"** is the fundamental, basic, and constituent ideas, philosophies, mathematics, and technologies which are necessary and were needed as prerequisites to develop the modern field now called **"Computer Science"**. If we define the prerequisite fundamentals needed for computer science as **Logic**, **Binary Mathematics**, **Semiconductors**, and **Computers**, then in order to prove the truth of the statement that "the fundamentals of computer science started in Africa", we would only have to show where logic, binary mathematics, semiconductors, and computers existed on the African continent prior to any place else on the planet. The following sections of this book are devoted to providing the empirical evidence and examples to substantiate and prove the aforementioned premise. Moreover, it can also be shown how other topics related to computer science such as Robotics, Artificial Intelligence, Virtual Reality, and Transhumanism originated on the African continent and are related to traditional African culture and philosophy.

This book grew out of a collection of notes and transcripts from lectures, classes, interviews, and presentations related to African Science, Mathematics, and Technology given by "**Prophessor A.C.E.**" (**African Creation Energy**) between the years of 2009 and 2012. The title and dates of these presentations are as follows:

- "**The African Origin of Binary Code**" a presentation from the book "**Supreme Mathematic, African Ma'at Magic**" in November 2009

- "**The Ontology of Virtual Reality and Artificial Intelligence**" a lecture and class by "**Prophessor A.C.E.**" in May 2010

- "**Black Light: The African Origin of the Illuminati and the Influence on Modern Science**" a lecture by "**Prophessor A.C.E.**" in October 2011

- "**Afro-Futuristic Consciousness – Breaking the Spell of Technophobia**" a presentation by "**Prophessor A.C.E.**" in March 2012

In all of these presentations, there was a theme related to Africa and Computer Science that was noticeable as if the evidence was pointing to an inevitable, reasonable, and unavoidable **logical conclusion**. As more and more research was done on traditional African culture, customs, sciences, and technologies, it was as if the data being downloaded into my brain from my African Ancestors was instructing me that a book on **African Computer Science** needed to be written. I concluded that the African Initialization of Computer Science must take place. However, upon arriving at this logical conclusion, I was somewhat uncertain about how the information would be received.

Many Africans and people of African descent view the computer and computer-based technologies as something "evil" and/or non-African. During my first visit to Africa in 2008, an elder in **Ghana** stated that the "**computer is the white-man's talisman**". When I asked the elder to elaborate on what he meant, he went on to tell me that in **West African Vodoun**, a **talisman** is a tool used by certain priests as an extension of the priest's power and can be used to **hypnotize** people into a **trance**. He further stated that the television and the computer are two forms of talismans that are used by white people to put other groups of people into a **hypnotic trance** and under the control of white people. While I comprehended the analogy that the elder was making between the "magic" of a talisman and the "magic" of computers, I also recognized that the computer was simply a tool, and like any tool, it could be used for productive and constructive intent, or it could be used for malicious and destructive intent. However, the sentiment of fear and contempt towards the computer that was expressed by the elder in Ghana is also shared by many people of African descent in the Diaspora. As a child and as a young adult growing up in "Afro-centric", "Conscious", and "Cultural" communities, I often heard stories of "**The Illuminati**" and how computer technology would be used in the future to enslave people by placing a **computerized microchip implant** inside the entire Human population which would serve as "**the Mark of the Beast**" in the "**New World Order**". The stories went on to explain how the microchip that would be placed inside of people would be capable of taking control of the person's brain and transform the person into a **mindless zombie** or "**Robot**" under the control of the "**Illuminati**". In addition to the microchip implants, the stories also explained how the Illuminati had a "**Robot Agenda**" in which they would use movies, videos, and other forms of media propaganda to get humans to have an affinity towards Robots and accept the idea of living with Robots and even becoming a

Robot. Part of the evidence used for proving the validity of these claims usually included references to popular **Science-fiction movies** and also analysis of the logos of various computer technology companies. A comparison between the "**Eye of Providence**" symbol over the incomplete **Pyramid** on the back of the **U.S. dollar bill** to the various logos of technology companies that included the eye motif, the pyramid motif, or both, would be used as evidence that the "Illuminati" was behind the invention of the computer, and that all computer technology was evil. These philosophies, ideas, and beliefs discourage many young people of the African Diaspora from wanting to learn computer science and technology, and this in turn leads to a racial disparity in achievement in the field of Computer Science. Ever since I was a child I have had a Natural gift for developing technology. For my **6TH** grade **Science Fair** I built a remote-controlled Robot. The Robot's eyes were made from two red lights, and the nose was a power knob taken off of a fan. When the Robot's "nose" knob was turned, the Robot would turn on, and the eyes would light-up. The feet of the Robot were the wheels from a Remote-controlled car, so once the Robot was turned on and the eyes lit-up, the Robot could roll around the room directed by my remote control. Although I won first-place in the science fair that year, after the competition was over, one of the elders at the school pulled me to the side and explained to me how my science fair project was working against the progression of Black people because it promoted **idolatry** and "**Devil worship.**" I was also told that my natural given talent at **drawing pictures** of people was a form of idolatry and a **sin**. And so from an early age, it was apparent to me that not only did racist individuals of other races see Computer Science and Robotics as fields that Africans and people of African descent should not participate in, but some Africans and people of African descent also believed that Computer Science and Robotics are fields that people of African descent should avoid. In my mind, this logic did not compute. If the oppressor and the oppressed should not share the same "God", then likewise, the oppressor and the oppressed should not share the same "Devil". For Africans in the Diaspora who are descendants of slaves, there was a point in time where we were discouraged and not "allowed" to read for fear that if we could read, then we would be empowered to do great things. Now it seems as if conspiracy theories have become a new form of mental shackles, and for prescribers and proponents of conspiracy theories, these ideas hold us back from making great accomplishments in the fields of Computer Science and Robotics. For Africans on the continent who experienced Colonization, or Africans in the Diaspora who experienced Slavery, much of traditional African culture, customs, cosmologies, philosophies, and ideas have been lost due to foreign influence. In my research and travels, as I delve deeper into African culture prior to foreign influence, I realized that most of the symbols, concepts, and ideas that we have been taught to fear as evil, were actually African symbols and philosophies that empowered our Ancient African empires and ancestors. One of the most powerful African symbols that we have been taught to fear as a symbol of "evil" and the "Illuminati" is the symbol of the **"Eye and Pyramid"** spoken of earlier. In Ancient Egyptian cosmology, "the eye" was a symbol for the **Sun deity RA** who, among other interpretations, represented **"Energy"**. The Pyramid was symbolic of "the **primordial mound of creation**" which, among other interpretations, represented **matter**. Therefore, the combination of "the eye and the pyramid" was symbolic of the scientific **Union** and **combination** of **"Energy and Matter"** ($E=mc^2$), or the creative process of how **"Thoughts become Things"**, the **ability to Create**, and the transformational **thermodynamic cycles in Nature**. The "eye" symbol is also depicted by the Ancient Egyptians who migrated to West African in the Adinkra symbol known as **Abode Santann** which represented the **union** of Natural and Human creations. It is a well know fact that

Pyramids originated in Africa. However, most of the Pyramids in the world are not in Egypt, but in the African country of Sudan.

There are 220+ Nubian Pyramids in Sudan built by the Kushite Empire, and most of these Nubian Pyramids were built in honor of **African Queens**. When the angle of inclination of Pyramids is examined, it can be shown that the Pyramid in the "Great Seal of the United States of America" that is said to represent the "Illuminati" is closer in design to a Nubian Pyramid than an Egyptian Pyramid. The "eye" or "light" or "energy" on the top of the Pyramid was also depicted in Ancient Egyptian iconography as a Bird on the top of the Pyramid. This bird was called the "**Bennu Bird**" which was the predecessor of the "**Phoenix Bird**" who

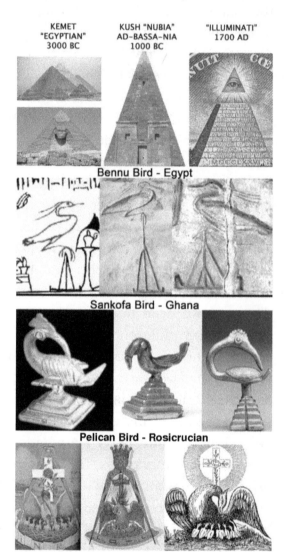

KEMET "EGYPTIAN" 3000 BC
KUSH "NUBIA" AD–BASSA–NIA 1000 BC
"ILLUMINATI" 1700 AD

Bennu Bird - Egypt

Sankofa Bird - Ghana

Pelican Bird - Rosicrucian

would destroy itself in "fire" (light, energy) and be reborn from the ashes. Before the "Phoenix Bird" was adopted into **Rosicrucian** "Illuminati" symbolism, it was used in the symbolism and philosophy of the West African **Mali** and **Ghana Empires** as the **Sankofa Bird** on top of a Pyramid. The Sankofa symbol, which means "**GO BACK AND GET IT**", is still used in West African Adinkra symbols today.

Eye of Horus from "Kemet" Ancient African Egyptian Cosmology

Abode Santann, West African Adinkra Symbol

The "Great Seal" of the United States, the "Eye of Providence". Said to be "Illuminati" Symbolism

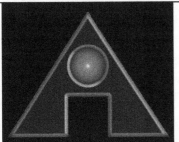

Logo for the 17th International Artificial Intelligence Conference

Logo for "America Online" Internet Service Provider

Logo for the "Second Life" Virtual Reality game

Transhumanism Logo

As a symbol of Science, **Creative Power**, and **Creative Energy**, it is not surprising that modern Technology companies, organizations, and movements such as "**America Online**", "**Second Life**", "**Transhumanism**", and "The International A.I. Conference" have all adopted versions of the African "eye and pyramid" motif for their logos. Proponents of "Illuminati" conspiracy theories would suggest that all of these technology groups who use this symbolism are being controlled by the "Illuminati" and are using computer technology to implement their nefarious schemes. For African people, it is important to recognize manifestations of your traditional African symbols and philosophies and utilize them to be empowered and take control whenever and wherever possible.

Believers in the "Illuminati" conspiracy theory also suggest that there is a "**Robot Agenda**" being implemented to get masses of people to accept the notion of becoming a "Robot". In recent years there has been a multitude of entertainers, musicians, actors, and celebrities that have utilized some form of robotics, androids, or cybernetics in their performances and this is cited by conspiracy theorist as evidence to prove the legitimacy of the "**Robot Agenda**" theory. The Conspiracy Theorist will go on to say how the oldest depiction of the "Robot Agenda" in film was a silent movie called "**Metropolis**" which depicted the creation of a female Robot, and behind the Robot was a large **inverted pentagram**. The Conspiracy Theorist will say that the inverted pentagram is a symbol of the "**Devil**, **Satan**, and **Lucifer**" and the creation of the female Robot in the Metropolis movie with the inverted pentagram behind her is evidence that Robots are evil. They will tell you that this inverted pentagram is symbolic of the "**Goat of Mendes**". They will even go on to tell you that the name "**Bill Gates**", the founder of the **Microsoft computer company**, is in reference to the "**Billy Goat**" and the "**Gates of Hell**" to further suggest that computers are evil.

However, when scholastic research is done into the topic of the **"Goat of Mendes"**, we find that Mendes was an ancient African city in **Kemet** or Egypt. Initially, the deity in the city of Mendes was a now extinct species of **Ram** with horizontal spiraling horns scientifically called *Ovis longipes palaeoaegyptiacus*, and when these Rams died out, the Ram was replaced with the **Billy Goat**. Rams are male **sheep**. Although Rams and Goats are both members of the scientific family called **Bovidae**, one of the factors which **separate the Ram from the Goat** is the symbolic use in Ancient African cosmology. The Ram

Scene from the movie Metropolis with the "inverted pentagram" behind the newly created female Robot.

The inverted pentagram is said to be symbolic of the "Goat of Mendes"

was venerated as a symbol of **Creativity** because of its virility for copulation and procreation. Humans still pay homage to the reproductive libido of Rams and Goats by taking herbs called *Epimedium* or "**Horny Goat Weed**" to try to achieve the same level of virility. In the city of Mendes, the Ram was called **Ba-Neb-Djed**, which meant "**soul of the lord of stability**". The word for "soul" and the word for "ram" in Ancient Egypt was called "**BA**". The Ram of Mendes, Ba-Neb-Djed, represented the soul or "**BA**" of the sun deity "**RA**" or the **RA-BA**. A stele in the Ramesseum tells the story of how the African Creation deity named **Ptah** manifested the virility of the Ba-Neb-Djed to impregnate the mother of the **Pharaoh Rameses II**. In southern Egypt, Sudan, Nubia, Kush, and Central Africa, the Ram Ba-Neb-Djed was equivalent to the deity **Khnum**. As an ancient and primordial deity worshipped from the Predynastic period of Ancient Egypt until today, Khnum was associated with the **craft** of **pottery** and was said to have created the **body** and "spirit" **KA** of the first human beings on his **potter's wheel**. Under the name **Khonvoum**, Khnum is still revered by the **Mbuti** (**Bambuti**) **Pygmy** people in central Africa today. It is said that Khonvoum created the different races of people on the planet from different types of clay with black people being created from black clay, white people being created from white clay, and the Pygmies being created from red clay. The name "Khnum" is derived from the Ancient Egyptian root word "***khnem***" meaning "**to join, to unite**, and **to build**". It is said that the attribute of the **Islamic deity Allah** of *Khāliq*, meaning "**Creator**" in **Arabic**, was derived from the name and story of the African Creation deity Khnum. In fact, one of the names of the "goat of mendes" is "**Baphomet**" which was derived from a mispronunciation of the name of the Islamic prophet **Muhammad** as "**Mahomet**". The Ram deity Khnum, Ba-Neb-Djed, or the "Goat of Mendes" was demonized by early Christians because of the sexual associations related to his mythology. Naturally, any other philosophies or ideas that seemed to challenge Christianity were demonized and associated with the "Goat of Mendes" as is the case with Islam and Muhammad (Mahomet/Baphomet) and even Science and Robotics. The concepts of the "Devil" and "evil" are religious and theological concepts; therefore ultimately anything that someone perceives as "evil" or the "devil" must have been established by some religious or theological doctrine. It is a well know fact that the **Theologies of Religion** have been at odds with the **Theories of Science** for hundreds of years with religion portraying Science and science related topics such as technology, computers, and robotics as "evil". "Natural Selection versus Intelligent Design", and "Evolution versus Creationism" have both been manifestations of the Science versus Religion paradigm. The book and movie entitled "**Angels & Demons**" also depicted the Science versus Religion conflict as the **Illuminati Scientists** versus the **Vatican Priests**. Thus, it is not surprising that religious dogma would try to demonize Robotics by placing the inverted pentagram symbolic of the "Goat of Mendes" (the Ram Ba-Neb-Djed or Khnum) behind the Robot in the movie *Metropolis* in an effort to portray Science as evil and associate it with what religious dogma believes is the "Devil". However, there is a surprising connection between the African Ram-Headed deity of Mendes named Khnum and the operative Science of Robotics. The act of Khnum **creating a being out of clay** on his potter's wheel and then **animating his creation with the energy called "KA"** is very similar to a Scientist or Technician **creating a Robot** and **animating the creation with electrical energy**. Moreover, in the "Hymn to Khnum" from the city of Yebu (also called Abu, Elephantine, or Esna), Khnum is equated to the "RA-BA" (the BA of RA) and also to the deity **Ptah** as **Ptah-Tatenen**, the **creator of creators**. In Ancient Egyptian cosmology, Ptah created the Heavens, the Earth and the Universe, and then Khnum created the people and living things in the Universe. Another way to interpret this metaphor is to say **Ptah Created the "Virtual Reality"**

(Universe) and **Khnum created the "Robots"**. When we combine these three attributes of Khnum, we get **RA-BA-PTAH** also pronounced **"RA-BA-TA"** which is phonetically similar to the word **"ROBOT"**. Also, the word "Ram", which is associated with Khnum, is also used in computer science as the abbreviation for "**Random Access Memory**" or **RAM**.

KHNUM - The "Ram of Mendes", the African Creation deity called the creator, potter, builder, and molder. Designer of the Human "Robot" in Ancient African Mythology

There have even been statements made by proponents of **Illuminati** conspiracy theory that suggest that since the word "**Robot**" is phonetically similar to the name "**Robert**", and the name "**Robert**" means "**bright**", then this is further evidence of the Illuminati's involvement in the field of robotics because the word "**Illuminati**" also refers to **light** and being "**bright**". The word "Illuminati" also means to be "**Enlightened**" or "**Conscious**". Therefore, it is a logical fallacy for someone to say they are "**enlightened**" or part of a "**conscious community**" and be against the "**Illuminati**". It is the Illuminati's reference to "**light**" which facilitated the Judeo-Christian religious association of the "**Illuminati**" to "**Lucifer**" because the word "**Lucifer**" means "**light-bearer**". However, there is no direct association between "**Lucifer**" to the "**Devil**" or "**Satan**" in the Judeo-Christian religious text. The word "Lucifer" only appears **once** in the Bible in the book of **Isaiah chapter 14 verse 12** where it states: *"How art thou fallen from heaven, O **Lucifer**, son of the morning star"*. In the Judeo-Christian religious text, there is more of a logical and reasonable connection based on "**light**" between the concepts of the "**Illuminati**", "**Lucifer**", "**Jesus**", and "**God**" than there is to the "**Devil**" and "**Satan**". In the Judeo-Christian Bible, both Jesus and God are equated to "**light**" and "**the bright morning star**" in the following verses:

- *"...I **Jesus** am the root and the offspring of David, and **the bright morning star**."* ~Revelation 22:16

- *"Then spoke **Jesus** again unto them, saying, I am the **light** of the world".* ~ John 8:12

- *"...declare that **God is light**, and in him is no darkness at all."* ~ I John 1:5

Thus, the attempt to demonize computer science and robotics by associating it with something "evil" is based on fallacious logic and unsound reasoning.

Another religious concept which attempts to induce fear into people about topics related to computer science is the "**Mark of the Beast**" or the "**Number of the Beast**" spoken of in Revelation chapter 13 verses 16 to 18 of the Judeo-Christian Bible where it states:

> 16. *And he causeth all, both small and great, rich and poor, free and bond, to receive **a mark** in their right hand, or in their foreheads*:

> 17. *And that no man might buy or sell, save he that had the mark, or the **name of the beast**, or the **number of his name***

> 18. *Here is wisdom* (**sophia**). *Let him that hath understanding* (**nous**) *count the number of the beast* (**thērion**): *for it is the number of a man* (**anthrōpos**); *and his number is Six hundred threescore and six* (**chi-xi-stigma**).

Conspiracy theorist have suggested that the bar codes that appear on items for purchase in retail stores somehow have the number 666 coded into them, and since the IBM computer company developed the software needed to read these barcodes, then IBM and computers in general must be "**the beast**" spoken of in the Bible. Conspiracy theorists go on to suggest that since the name **HAL**, which was the **Artificial Intelligence** program in the movie "**2001: A Space Odyssey**", is a one letter shift from the letters "**IBM**" (**H**-<u>I</u>, **A**-<u>B</u>, **L**-<u>M</u>), then **Artificial Intelligence** is also evil. The book "**Supreme Mathematics, African Ma'at Magic**" by **African Creation Energy** explains how Bar Codes can be generated by vertically extending the dots-and-dashes of **Morse code**. Moreover, Morse code and **Bar Codes** are **Binary** methods of expressing letters, words, and numbers just like the beat of a "**talking drum**" which all have their origins in Africa and will be expounded upon more in the following chapters of this text.

The methods of encoding information in "bar codes" have been extended into **2-Dimensional Matrix Bar Codes** and **3-Dimensional Matrix Bar Codes** to develop what is known as QR (Quick Response) Codes. The 2D **QR Codes** can be used to encode a variety of different types of

2D Matrix QR Code for the website http://africancreationenergy.com

data including websites and phone numbers. The 2D QR codes can be scanned with the camera feature on smart-phones and other mobile devices as an interface between the cyber world and the real world. 2D QR codes have also been generated for an individual's personal website or online profile, and thus if a person is wearing an item that has a QR code for their personal website or online profile, you could scan the QR code of a stranger and be directed to their personal profile. In this case, it is interesting to note how much 2D QR codes look like **fingerprints**.

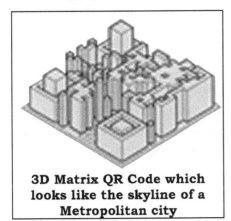

3D Matrix QR Code which looks like the skyline of a Metropolitan city

Where **2D Matrix Barcodes** can contain more information than 1-Dimensional barcodes, **3D Matrix Barcodes** can encode even more information than 2D Matrix Barcodes. 3D Matrix Barcodes look like the **skyscraper buildings** of various heights in a metropolitan city. Potentially, a city could be constructed in the layout of a 3D Matrix barcode to encode data that could be scanned and read by Airplanes or various **interstellar crafts** passing overhead. Therefore, you could be living in a metropolitan city, and <u>actually</u> be **living in "the Matrix**" or 3D QR-Code. If conspiracy theorist believe that barcodes are the "**number of the beast**", and that the "number of the beast" is a 1-Dimensional **1-by-3 Matrix** [6 6 6], then they will need a new religious text so they can create new conspiracy theories to account for higher order matrices like a

2-by-3 Matrix $\begin{bmatrix} 6 & 6 & 6 \\ 6 & 6 & 6 \end{bmatrix}$ in order to keep up with technology

in an attempt to demonized 2-Dimensional Matrix QR-Codes.

First of all, the Bible clearly says that the "**number of the beast**" is the "**number of a man**" or human, not a machine, computer, or Robot. Therefore, it is a logical fallacy to conclude that the "number of the beast", 666, is somehow related to computers or robotics. Religionist will attempt to demonize the concept of **Robotics**; however, Religionist will tell you that "**Angles**" are good, but "Angels" in religion are identical to "**Robots**" with no **free-will** and mandated to follow the instructions of "God". Conversely, in Religious theology, Human beings with free-will have the ability to disobey the instructions of "God" and do acts of "evil". Perhaps what Human beings truly fear the most is that the Robot may become like the Human, with free-will and the ability to "**know good from evil**" and **disobey**.

Secondly, lets us specify what qualifies something as **Organic** and what qualifies something as **Inorganic**. Although currently there is debate between Biologists and Chemists as to what determines if something is Organic or Inorganic, Chemist define "**Organic**" as any molecule or compound that contains the element of **Carbon**. Therefore, "**Inorganic**" would be the molecules or compounds that **do not contain Carbon**. This is why when scientists are searching for "Organic" life on other planets; they search for "**Carbon-based life forms**". The most abundant stable isotope of Carbon is **Carbon-12** which has <u>**6**</u> protons, <u>**6**</u> neutrons, and <u>**6**</u> electrons (**666**) and accounts for 98.89% of all Carbon. Therefore, if you believe that **666** is indeed the "**number of the beast**", then it would be **Organic** things which contain Carbon-12 with 6-Protons, 6-Neutrons, and 6-Electrons or 666 which would be "Evil" and "of the Devil"; Conversely, Inorganic things like computers, robots, and machines which do not contain Carbon-12 and the number 666 would be "Good" and "of God". Moreover, if you believe that "Organic" things are "good" and "Inorganic" things are somehow evil, then following our discussion which showed that Organic means "the presence of Carbon" and Inorganic means "the lack of Carbon", then if a Robot, Computer, or Machine is made with a compound that contained Carbon, then that Robot, Computer, or Machine would be "Organic" by definition. **Steel** is a metal alloy made by combining **Iron**, **Carbon**, and other elements. Therefore, a Robot made with Steel would be an "**Organic Robot**" by definition.

The point is that much of the fear mongering and attempts to demonize computers, science, and technology are based on fallacious logic and unsound reason. This irrational fear of advanced technology, science, and computers is called **Technophobia**. In Western culture, Technophobic propaganda was perpetuated during the era of **Romanticism** which was a movement in 18th century Europe which started as a reaction of revolt against the **Industrial Revolution**, the **Age of Enlightenment** (**Age of Illumination**), Science, and Reason. Romantics tended to believe in **imagination over reason** and the **organic over the mechanical**. Romantic Technophobic ideas have manifested in a number of arguments and debates over the years including **Religion versus Science**, the Vatican Church versus the Illuminati, Creationism versus Evolution, and **Natural Selection versus Artificial Selection (Eugenics)**. The **Amish Mennonite** resistance to using technology is one manifestation of Romantic Technophobic philosophies. Most technophobic ideas stem from people trying to hold on to old philosophies and theories that have been disproven. For instance, the root of the **"Organic versus Inorganic"** debate that was previously discussed stems from an ideology called *Vitalism* from around 100 AD. The doctrine of "Vitalism" made a distinction between "living" and "non-living" things and suggested that "living things" cannot be created from "non-living" things. The doctrine of Vitalism was discredited when modern scientists were able to synthesize complex organic compounds from inorganic substances. However, even after empirical evidence and experiments proved the doctrine of Vitalism wrong, proponents of the doctrine still had a desire to separate, and thus we have a modern day debate between Biologists and Chemists about what is "Organic" and what is "Inorganic". Also rooted to the doctrine of Vitalism which seeks to separate "living" things from the rest of the world and separate Humanity from the rest of Nature are debates about **"Natural versus Synthetic"** and **"Natural versus Artificial"**. First of all, to suggest that something is "Not Natural" is to suggest that something can exist outside of Nature. If it exists and has manifested in Nature, than it is Natural. In regards to the **"Natural versus Synthetic"** arguments and debate, although many Thesauruses list the word **"Unnatural"** as a synonym to the word **"Synthesis"**, the opposite of the word **"Synthesis"** or **"Synthetic"** is not the word "Natural" but rather the word **"Analysis"** or **"Analytic"**. The etymology of the word **"Synthesis"** is **"to put together or combine."** Therefore, the word **"Analytic"** from **"Analysis"** meaning **"to break apart"** would be the opposite of synthetic. Synthetic and Analytic are comparable to the words **Construction** and **Destruction** respectively. Thus, comprehending the meaning of the word "Synthesis", we can logically say that every created thing in Nature and the universe is synthetic in that it is the combination of the fundamental particles of Nature.

In regards to the **"Natural versus Artificial"** debate, the word **"Artificial"** means **"something made by Human workmanship, craft or skill"** and the word **"Natural"** means **"of or relating to**

Nature". African philosophy views the Human being as a part of nature, and subsequently anything that a Human being creates is also a part of Nature. A **Bird** is a part of Nature. If a Bird builds a **Nest**, is the Nest Unnatural? A **Bumblebee** is a part of Nature. If a Bumblebee builds a **Bee-Hive**, is the Bee-Hive Unnatural? A **Beaver** is a part of Nature. If a Beaver builds a **Beaver-Dam**, is the construction Unnatural? **Ants**, **Termites**, and **Moles** are part of Nature. If Ants, Termites, or Moles build **mounds**, are the mounds Unnatural? So then if **Humans** are a part of Nature, why would human creations like computers, robots, or machines be considered unnatural? The source material for all human creations and inventions can be traced back to Nature. The **unification** of Human and Natural creation in the **totality of the universe** is expressed in the West African Adinkra symbol called "***Abode Santann***". The Abode Santann symbol depicts 3 forms of creation in one symbol: The **Sun**, the **Moon**, and the Stool (Human or **Earth**), *Re-Aah-Ptah.*

Abode Santann complete symbol	**The Sun (Eye)**	**The Crescent Moon**	**Stool**

The book "**The Science of Sciences and the Science in Sciences**" by **African Creation Energy** describes the symbolic significance of the Sun and the Moon in Ancient African philosophy. The **Sun** is said to represent the **Empirical world** and the **Moon** is said to represent the **Rational world**, with the **Human** capable of unifying the two worlds. One application of this African philosophy is in the world of **computers** with **Hardware** representing the **empirical world (the Sun)**, **Software** representing the **rational world (the Moon)** and the **User** (the Human or Earth) as the unifier of the two dualities through creation. The step-by-step process of unifying these two dualities to accomplish goals or solve problems is called an **Algorithm**. When the approach to **goal accomplishment**, **unification**, and **problem solving** is based on **evidence**, **experience**, and **experimentation** it is called a "**Heuristic**" coming from a word meaning "**to discover or find**".

The word "**Heuristic**" is also phonetically similar to the word "**Heru**", the name of the Ancient Egyptian deity who lost an "**eye**" in his battle to avenge his father **Osiris**. Therefore, it is not surprising that the **Artificial Intelligence computer** from the movie "*2001: A Space Odyssey*" named **H.A.L. 9000** which means "<u>H</u>euristically programmed <u>AL</u>gorithmic computer <u>9000</u>" was portrayed by the symbol of an eye.

HAL 9000
Heuristically programmed Algorithmic computer 9000 from the movie "2001: A Space Odyssey"

There are a multitude of other symbols and concepts that we have been taught to fear as symbols of "evil" or the "Illuminati" that when the origin of the symbols are traced, you find that the symbols are derived from traditional African culture, philosophy, and theology. For example, it is said that "the **Spider's Web**" is symbol of "**the Devil**" and the "**Illuminati**" and thus the **internet** was named the "**World Wide Web**" because it is being controlled by the Illuminati and forces of evil. However, to the **Akan** people of **West Africa**, the "**Spider's Web**" represents the web of **Anansi the Spider** which permeates all of existence and is comparable to the modern concept of **String Theory**. It is also said that the "**Owl**" is an evil symbol of the "**Illuminati**" which stands for "**Order of the White Lodge**", however the **Owl** was revered in Africa and can be found in the **Hieroglyphs** from **Ancient Egypt** in the name of the Great **African**

Architect who built the first **Pyramid** named **Imhotep**. Even the geometry of the **"compass and square"** symbol used in the **Freemason** secret society which is often linked with the **Illuminati** has its origin as a symbol used for African/Haitian deity **Obatala / Damballah.**

ANANSE NTONTAN the "spider's web", West African Adinkra symbol of wisdom and creativity 	The Devil's Web 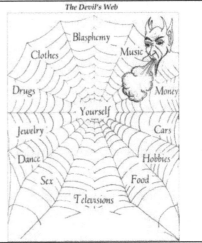
African Hieroglyphics, (Medu Neter) for the Name IMHOTEP containing the symbol of an Owl (letter "M") 	Symbol of the Bohemian Club "Illuminati" group containing an Owl and the message "Weaving Spiders come not here" 
Symbol for the Haitian/African deity Damballah/Obatala 	Compass and Square of the Freemason "Illuminati" group

The purpose of this conversation is not to legitimize the Illuminati or to try to portray the Illuminati in a positive light. The purpose of this discussion is to show African people that many of the symbols and concepts that you have been taught to fear as something "evil" have been symbols and concepts that you once used when you ruled the world. Hence, **you have been taught to fear yourself.** The purpose of this discussion is not to get African people to participate in the utilization of technologies and inventions created and developed by other groups and races of people by become "technological slaves" to the detriment of Africa. NO! The purpose of this discussion is to get more African people to become the **creators and inventors** of highly advanced forms of technology that can then be sold and exported to Africans and other groups and races of people around the world for the benefit of Africa.

What has been shown thus far is that another group of people has utilized the culture, symbols, cosmologies, and philosophies of African people to empower themselves and simultaneously taught African people to fear it. Also, many of these symbols and cosmologies have been utilized in fields related to **Computer Science.** Historically, the **"Illuminati"** refers to several different secret societies that were developed in the 1600s and 1700s during the **Enlightenment** era or **"Age of Reason"** in Europe. One of these Enlightenment era groups was the **Rosicrucian** organization. The ideas of Rosicrucian philosophy gave birth to an organization of **12 Natural Philosophers** or **"Scientists"** based on the works of **Sir Francis Bacon** (*"Jesus and his 12 Disciples"*) called the **"Invisible College".** The Invisible College went on to become the **"Royal Society",** which is the oldest society dedicated to Science in the Western and European world. The **Royal Society** has influenced scientific research, and many of the

Scientists that have shaped the Science and Technology of our modern world have been members of the Royal Society. Many Scientific theories and technologies are the result of **Occult** and **Esoteric** concepts that can be found in traditional **African Theology** and Philosophy.

African people, **now is the time** to accept who and what you are and apply your traditional African culture, philosophy, and cosmologies. Other groups of people have embraced your African culture, and it is empowering them, but you refuse to accept who and what you are. **Actions speak louder than words!** If African people reject their traditional African culture and philosophy and embrace something other than their traditional African culture and philosophy, their actions are saying that they love other than self and kind, and are thankful for their slavery and colonization. You deny Egypt, you deny Sudan, you deny the Pyramids, you deny the "eye", you deny "the Ram", but the people who are creating the technology, computers, hardware, and software that you fear are embracing all of these things. If you believe the "Illuminati" are in Power and their **"Black Magic Rituals"** is what enabled them to get in power and stay in power, then doesn't that validate the fact that **"Black Magic"** rituals actually work?! And, if you are not in power, and you are not doing **"Black Magic Rituals"** then doesn't that prove that what you are doing does not work?! If you believe in **"The Mark of the Beast"**, this presupposes that you believe in **Christian** and/or **Islamic doctrine**, which means that if you are a person of African descent, you agree and are happy that white people or Arab people removed you from your traditional African religion and culture and gave you a new religion to inform you that there was a "Beast" in the first place.

The "God" concept has been programmed into your mind by foreign Religion to create a mentality that someone or something is going to come and save you from your problems, or solve all your problems for you. The "Devil" concept has been programmed into your mind by foreign Religion to create a mentality which you can blame all your problems on as a scapegoat and not accept responsibility. Amongst the **"Conscious Community"** of Africans on the continent and in the Diaspora, many teachers have come to **re-program** your mind and remove the **"God"** concept that has been given to you by foreign Religion, and therefore now you must save yourself and solve your problems for yourself. Now, the **"Devil"** concept that has been programmed into your mind by foreign Religion is being taken from you so that you no longer have someone or something to **blame** for your problems and shortcomings and not accept responsibility and accountability.

With the mental shackles of the "God" concept and the "Devil" concept removed from the African mind, what will be left is **YOU**, and you will have no choice but to **CREATE** all that is needed for the survival and well-being of your family and your kindred. YOU will manifest all of the extraordinary abilities that you once placed outside yourself which you either feared or revered. You will receive actual ETERNAL life, for as long as you want it, if you return to your traditional African culture and take part in the operative expression of your traditional African philosophies and customs. TAKE HEED, for Judgment Day approaches, and you have to decide if you will take control of your reality or if you will let your reality control you.

People fear what they cannot control, and they cannot control what they do not comprehend. Therefore, the first step to getting over your fear is to study whatever it is that you are afraid of so that you can comprehend it. The second step is to apply the knowledge that you have learned while studying whatever it is you are afraid of, and control that which you fear. The "application of knowledge" is the meaning of the word Technology. Thus, **every living thing in Nature uses some form of Technology**; that is to say, every living thing in Nature applies some form of knowledge or information. A plant that adjusts its leaves to face toward the light is utilizing technology or "applying knowledge". Animals that go out to hunt for food at certain times, or only eat certain foods and do not eat other foods are utilizing technology or "applying knowledge". It is the application of knowledge, or Technology, which enables you to take control of your reality. Moreover, the reality about the concept of "God" is that it can be realized and achieved by Humans through computer technology. Let us consider the four characteristics of the "God" concept as being omnipresence, omniscient, omnipotent, and eternal. Then the **Omnipresent** (meaning **ubiquity** or "to be everywhere") aspect of God can be accomplished by a form of computer technology called **Ubiquitous Computing** where computer technology, cameras, microphones, and other sensors will be integrated into everything everywhere including buildings, furniture, cars, clothes, shoes, etc. The ability to be everywhere facilitates the ability to gain all knowledge and know everything and achieve another aspect of the "God" concept called **Omniscience** meaning "all knowing". The ability to use and apply all knowledge facilitates another aspect of "God" by becoming **Omnipotent** meaning "all powerful" or "all mighty". Lastly, the ability to transfer one's consciousness into another form like a Robot facilitates the ability to transcend death and realize the fourth attribute of "God" and become **Eternal** and **Immortal**. Thus, at this point in time, computer science and its related technologies is a path which could enable us to "**wear the mask of god**". The etymology of the word "**Satan**" is "**adversary** or to **oppose** or **go against**". Thus, anything or any **problem** that prevents you from accomplishing something is a "**Satan**" by definition. Do not let your **fear** be your "**Satan**" and get in your way of becoming "**God**" by not learning to create and master **computer science** and **computer technology**.

A good scientist and technician will consider both the positive and negative outcomes of an action. With computer science and its related technologies, there are many positive possible outcomes, but there are also some negative potential outcomes. For African people, we have been told about the negatives and discouraged from science and technology more than we have been told about the positives and encouraged in the creation, invention, and development of African sciences and African technologies. The reality about technology is that a "Brain Child" in the form of an invention is just like a child or offspring in that it takes on certain characteristics and traits of the inventor. Thus, the computer science and technology created and invented by individuals who have a disdain towards African people could likely be detrimental to African people. If you prescribe to the belief that negative dystopian events related to computer science and technology will occur, you must also agree that comprehending the fundamentals of computer science, robotics, and programming will empower you with the ability to combat and override anything that an adversary may be planning. If your adversary is planning to implant a computer chip inside of you, you must know computer science in order to re-program or deactivate the chip. If your adversary is planning on embarking on **Cyber-warfare**, you must know computer science in order to protect and defend yourself against a **Cyber-attack**. If your adversary is training Cyber-Soldiers, you must know computer science in order to create and develop **African Revolutionary Cyber-**

Soldiers. If your adversary is developing self-steering "smart bullets" that can be programmed to turn corners and "seek and destroy" a target, then you must know computer science in order to develop a self-steering "smart" bullet defense system which will be able to shoot the incoming "smart" bullets out of the air much like a **"missile defense system."** And so the study of computer science and computer technology is paramount for any Revolutionary in the present and in the near and coming future.

The reality is that Technology is not inherently good or bad; it is the intention and use of technology that can be either good or bad. **Fire** can be used to **provide warmth** in cold climates or it can be used to by an **arsonist** with evil intent to burn down someone's house. However, it is not the fire that it bad, it is the wielder of the fire that is bad. Guns don't kill people, People do. Cars don't kill people, People do. Even Technology and tools that are used for Agriculture can be used for evil intent. The weapon

Crook (Sickle) and Flail – tools of Agriculture from Ancient Egypt

Nunchucks weapon

called **"Nunchucks"** evolved from the agricultural Flail tool used for harvesting crops. War sickles and swords evolved from agricultural sickles. War maces and War Hammers evolved from the **blacksmith's hammer.** This is why in the West African **Yoruba** culture, the **Orisha Ogun** (also called **Yurugu** or **YAA-OGO** in **Dogon** cosmology) is a deity associated with both **Technology** and **War.** As agricultural tools, all of these technologies and tools were considered prestigious amongst the royalty in **Ancient Nile Valley culture**. These tools and technologies were also honored amongst the warriors in Ancient Nile Valley culture. Many of these tools and technologies that are considered "bad" when used for war actually evolved from tools and technologies used for agriculture and were considered "good" when used in times of peace. It is a logical fallacy to argue that just because something can be used with negative intent then that means the thing is negative. A spoon is a technology that was created with the positive purpose and intention to be used as a feeding tool, but it can also be used as a blunt force object to strike someone with out of anger. Does this then make the spoon itself negative? Computers and Robotics are technologies and tools that can be utilized positively or negatively, but that does not make the computer or the robot itself positive or negative.

Writing this book I potentially could be attacked by people on all sides. I could be attacked by non-African people of other races who desire to completely write the contributions of Africans out of history books, and cannot fathom that African people actually made contributions to the growth and development of the highly advanced field of computer science. I could be attacked by my own African brothers and sisters who have prescribed to the philosophy that computer science is a field of "the white man" and that I am somehow helping to promote an "Illuminati Robot Agenda" to manipulate and control African people. Let me be clear: Any form of technology (not just computers, robotics, Artificial Intelligence, or Virtual Reality) that is comprehended, controlled, and created by one group of people and is not comprehended, not controlled, and not created by another group of people who use it, can and will be used to enslave and exploit the latter. Regardless if it is Agriculture, Architecture, Automotive Design, or Fashion Design, if African people are not controlling and creating it, African people are subject to be under the thumb of those who are. It has always been the mission, goal, and aim of **African Creation Energy** to empower African people to take part in the creation and development of any form of technology which they utilize. Since it is indeed a fact that African people currently use computers, cell phones, and video games, and will likely use robotics, artificial intelligence, and virtual reality devices in the future, it is paramount that African people be the **Creators**, **Fashioners**, and **Makers** of all that they use, less they be further exploited and disenfranchised.

When African people are discouraged from participating, learning, and working in the field of computer science, it creates another area where African people can be potentially exploited. **Video Games** are a form of media based on computer science that will require more **African computer programmers** and **African computer graphic designers** to participate in the creation and development of Video Games which reflect African culture and values and portray positive images of African people. Since many video games are based on Comic Books and Cartoons, there are a few images of African people which appear in Video Games, Comic Books and Cartoons. However, many of these African characters that appear in these various forms of media were not created by or controlled by African people. Some of the portrayals in **popular fiction** of Africans and people of African descent with high-tech abilities and intelligence in the field of computer science include:

- **Dr. John Henry Irons** from the "**Steel**" Marvel comic book series (portrayed by athlete **Shaquille O'Neal** in the motion picture movie) who is an **Engineer** and becomes a new **Superman** by creating a **Robotic exoskeleton armor** to fight crime

- The **Cyborg Victor Stone** from the DC comic book series "**Teen Titans**" who is **half-man and half-machine** and has superhuman abilities due to cybernetic prosthetics he received in a near death experience

- **Major Jackson "Jax" Briggs** from the **Mortal Kombat** video game series and movies is the Major of a Special Forces Militia who has **cybernetic robotic bionic arms** and eventually becomes a complete cyborg.

- **Dr. Miles Bennett Dyson** (portrayed by actor **Joe Morton** in the motion picture movie) from the **Terminator** movie series was the inventor of an **evolutionary type of Microprocessor** which facilitated the creation of the **self-aware Artificial Intelligence** system called **Skynet**

- The Android **B1-66ER** from the "**MATRIX**" movie franchise was the Android responsible for initiating the **conflict between "man and machine"**. B1-66ER was threatened by his owner and eventually kills his owner. B1-66ER goes to trial and it is ruled that **machines do not have the same rights as humans**. B1-66ER is destroyed and the robots and machines riot in what is called the "**Million Machine March**". The robots and machines eventually create **their own industrial city in the Middle East** called "**Zero One**" which is very productive and threatens the **economic output** of the other human countries in the world. As a result of the economic disparity, a war ensues and the human countries devise a scheme called "**Operation Dark Storm**" which darkens the sky and **cuts the robots and machines off from the sun** as their primary energy source. Once the sun is blocked, the robots and machines are forced to harvest the energy from humans which leads to the creation of the "MATRIX" virtual reality simulation. Once the "MATRIX" simulation is created, "**The Oracle**" is a program within the simulation with "**predictive power**" and is portrayed in the image of an African-America woman.

The events which occur in the "MATRIX" movie franchise between **"Man and Machine"** are obvious allusions to conflicts which occurred between people of African descent and people of European descent in America. The android named **B1-66ER** is a reference to the character named **"BIGGER Thomas"** from a novel entitled **"*Native Son*"**. The court ruling that "machines do not have the same rights as humans" is in reference to the **"Jim Crow"** laws of **segregation** that once existed in America between Blacks and Whites. The machines organizing a "Million Machine March" is in reference to the **"Million Man March"** organized by the **Nation of Islam** and **NAACP** (National Association for the Advancement of Colored People) in 1995 as a demonstration to bring awareness to socio-economic discrepancies which exist in America between Blacks and white. And the act of the Humans attempting to "cut the machines off from the sun" was metaphysically symbolic since people of African descent are **"people of the Sun"**. These science fiction stories are based so much on the experiences of people of African descent in America that even the robots and machines in the science fiction story make the same illogical and unreasonable decisions that people of African descent have made despite the fact that robots are supposed to run on logic and reason. If people of African descent are being oppressed and mistreated in a place, rather than staying in that place and protesting, demonstrating, and begging the oppressor to accept and respect us, we should just leave. If the Robots in the movie were truly using logic and reason, rather than fighting and begging Humans to accept and respect them, they should have just left the planet and colonized the Moon, Mars, Venus, and any other planet where Humans could not go. In order for our stories to reflect our culture and our true mentality, Africans and people of African descent in our right mind must write our own stories. Africans cannot rely on others to create images of African people in video games, movies, or science fiction stories; we must do it.

As Africans and people of African descent participate in our traditional African culture and begin taking part in the operative expression our traditional African philosophies by studying computer science and developing computer software and robotics, we simultaneously create a new arena of employment and financial opportunity for our people. One of the biggest fears that people have about computers and robotics is that many of the labor-intensive jobs that humans have done historically can currently be done faster and more efficiently by computers and robotics. A popular depiction of this fear of robots and machines taking the jobs of people is depicted in the **African-American** folk tale of **John Henry.** In the story, John Henry was called the "**steel driving man**" who competed against a **steam-powered hammer** to see if he could lay a railroad track faster than a machine. John Henry won the competition against the machine, but then **died** at the end of the race with his **hammer in his hand.** The moral of the story that Africans and people of African descent should take away from the story of **John Henry** is to **work smarter, not harder.** It is a fact that many of the physically strenuous labor intensive jobs will be replaced by robots in the near future. It has already been demonstrated that **Autonomous Flying Robot helicopters** and **Nano-Quadrotors** are the "**builders of tomorrow**" and can be used to **build towers.** Also, the **Tiger Stone brick-laying machine** may eventually be developed to the point where human **brick masons** are no longer needed. However, there will still be a need to design, develop, build, program, and maintain these robots. So, while labor intensive jobs will be replaced by robots, new jobs will be created. In the United States of America, at the end of slavery, there was concern amongst some White American citizens that the freed slaves would "take their jobs" and therefore some White Americans argued against the freedom of Black Americans based on that illogical premise of fear; Likewise, it is just as unreasonable for people to use fear as a basis to argue against the creation and development of computer software, robotics, and machines. Therefore, studying computer science, technology, and robotics is a way to ensure job stability of Africans and people of African descent in the future. Computer Programming will have to be taught from Elementary level as it will be as necessary a language to know as one's spoken language.

If we consider that **secret societies**, like the **Freemasons**, grew out of the guilds of **operative** brick **masons** in the past, then it is likely that the secret societies of the future will grow out of the guilds of **computer programmers** and electrical and mechanical engineers responsible for developing the software and hardware of the **technologically advanced future**. Instead of symbols related to brick-masonry, construction, and building, the secret societies of the future that grow out of the teams of computer programmers, may use symbols related to computer science, robotics, and programming.

It is quite possible that even the participants in sporting events will be replaced by robots in the future. In the future, your favorite football, basketball, soccer, or baseball team may consist partially or entirely of robots. Fighting sporting events like boxing and martial arts may be replaced by robots in the future. Even artists in the music industry have the potential to be replaced by robots and machines in the future. Computers can be programmed to create music, and simulate lyrics, and holograms or robots can be used to display live performances. In 2012, the rapper and hip-hop artist **Tupac**, who had been deceased for 16 years, was **resurrected** in the form of a **hologram** that was programmed with the movements and the voice of Tupac at a music concert in **Coachella, California**. Even actors in movies have the potential to be replaced by computer simulated images in the future. The 2009 movie entitled *"Avatar"* was the first movie to earn more than **$2 Billion dollars** and many of the scenes in the movie were

Computer Generated Images (CGI). Currently, there are many Africans and people of African descent who have obtained great wealth and fame as professional athletes in sporting events or as entertainers, actors, or musicians. However, if and when the paradigm shift occurs and these fields become dominated by computers, robots and/or machines, all of the wealth that was once earned by Africans and people of African descent in these areas will go to the people who are programming the computers, and creating the robots and machines which will supplant the humans. In order to ensure that Africans and people of African descent maintain the wealth that has been obtained in athletics and entertainment, it is import for Africans and people of African descent who have obtained this wealth to invest in and sponsor computer and technology companies owned and operated by Africans and people of African descent.

Silicon Valley is being called "The New Financial and Economic **Promised Land**". Silicon Valley is the home to many of the world's largest technology companies, high-tech businesses, and silicon computer chip manufacturers. In July 2012, a **Ugandan Information Technology** (IT) student named **Abdu Sekalala** made a **fortune of Billions** designing **mobile phone Apps**. Some other Africans and people of African descent in the fields of Information Technology, Computer science, and Computer Technology include:

- **John Henry Thompson** comes from an **Afro-Centric Family** and has a degree from **MIT** in Computer Science. John Henry Thompson taught himself 5 different Computer Programming Languages and he even invented his own Computer Programming Language called *"Lingo programming"*, a scripting language which is used to create computer graphics.

- **Philip Emeagwali** is a native of **Nigeria** and an engineer and computer scientist who won an award from IEEE for his use of **supercomputing**. He has a bachelor's degree in **mathematics**, a Master's degree in **Engineering**, a Master's degree in **Applied Mathematics**, and a Ph.D. in **scientific computing**. He is considered the **Inventor of the World's Fastest Computer** and one of the "**Father's of the Internet**". Emeagwali was voted the "35th-greatest African (and greatest African scientist) of all time" in a survey by New African magazine.

- **Bertin Nahum** is a native of **Benin West Africa** and CEO of **Medtech**, a company which specializes in **robotic surgical equipment**. In 2010, Bertin Nahum created the **ROSA robot** which helps surgeons performs brain surgery. Bertin Nahum is considered the 4th most revolutionary high-tech entrepreneur in the world after Steve Jobs, Mark Zuckerberg and James Cameron.

- **Herman Chinery-Hesse** is a **Ghanaian Pan-African** and is considered the "**father of the African Technological Revolution**". He founded the **SOFTtribe computer software company** which is one of the largest and most successful software companies in West Africa that creates computer solutions for businesses across the African continent.

- **Valerie Thomas** is the Inventor and owns the patent to the *Illusion Transmitter* (U.S. patent #4,229,761 on October 21, 1980). The Illusion Transmitter is a technology which projects **3-Dimensional holographic images**. Valerie Thomas also worked on data systems for NASA.

- **Vérone Mankou** from the **Republic of the Congo** developed the **Way-C Tablet** and **Elikia smartphone** which is being called Africa's answer to the iPad and iPhone.

It is true that we can find on the African continent the origin and oldest examples of Mathematics, Analog Computing, and Logic, which are the fundamental and constituent parts prerequisite for Computer Science. However, just because you have all the parts of a machine does not mean you have the fully synthesized machine. A question to consider is "why and how is it that the fundamental concepts, philosophies, and tools necessary for computer science be found on the African continent prior to any other place on the planet, but the various components not be combined and synthesized by African to develop Computer Science"? Perhaps one answer to this question is that the many foreign invasions, migrations, colonization, slavery, and general exploitation which occurred across the continent over time prohibited Africa from synthesizing its components to develop modern Computer Science. However, since we can find evidence of the fundamentals of computer science in traditional African culture, philosophy, and customs, then these fundamentals can be used as great motivators for African people to emerge as leaders in the forefront of Computer Science and future technological creation and development. But before African people can begin creating and developing in the field of Computer science, it is paramount that we remove the psychological fears and "spells" related to computer science that have been programmed into our mind by colonization, invasion, and slavery. The preceding pages of this book were written to serve the purpose of "**Breaking the Spell of Technophobia**" as it related to computer science. If the old saying "**Where there is a Will, there is a way**" is true, then if people are unwilling to do something because they **fear** it or **do not see how it relates to them**, then there is no way the people will be willing to realize the objective. Therefore it was necessary to destroy your fears related to computer science, as well as give you examples of African computer scientists in fact and fiction to restore your **Creative Force of Will**.

Many firewalls on websites and computer systems have been configured to block **IP Addresses** from all over **West Africa** claiming that they are attempting to thwart what has become known as **"Nigerian 4-1-9 Advance Fee email scams"**. The result of having so many West African IP

Young men in Nigeria, West Africa accessing an internet hot-spot on their laptop Computers

Addresses blocked is that now internet access is intermittent in many areas throughout West Africa. However, it is not known if all of these email scams actually originated from the country of Nigeria, and it definitely does not make sense to block all West African IP addresses if it is know that the emails originated only in Nigeria. Thus, the act of blocking West African IP addresses is illogical and could be a form of **cyber-segregation**.

Therefore, there is an immediate need for African IT technicians and African Computer scientists to improve the internet infrastructure across Africa. Also, it is expected that the development of **Personal Robots** will be the next big **emerging technology** in the coming years which could produce the world's next multi-billionaire like the PC (Personal Computer) did in the 1980s and 1990s for **Bill Gates** of **Windows Computers**, and **Steve Jobs** of **Apple Computers**. Thus, Computer Science is a field which is based on African Philosophies, which African people could utilize to improve our economic paradigm around the world. The development of **African Computers**, **African Software companies**, **African Internet Service Providers**, **African Video Games**, and **African Robotics** are all areas which have a positive outlook for potential growth in the future.

And while the African country of **Nigeria** has achieved an **Aerospace program** with the **NigComSat-1 Nigerian communication satellite** as of May 13, 2007 and **automotive development** with the **Izuogu Z-600** car, the development of African computer science will lead to the growth, expansion, and improvement of the **African Aerospace** and **African Automotive** Industries.

Some critics may disagree with the term "**African Science**" or "**African Technology**", however for years, companies like **Mercedes Benz** and **Volkswagen** have been able to use terms like "**German Engineering**" unchallenged to instill a since of racial and national pride in their people and impress upon the world the German engineering prowess. Likewise, it is necessary to develop "**African Science**" and "**African Technology**" for the benefit of **African people**.

In computer programming, **initialization** is the assignment of an **initial value** for a data object or variable. The purpose of this book is **binary**; that is to say **two-fold**. One purpose of this book is to show how the fundamentals of Computer Science were "Initialized", originated, and began on the African continent. The following chapters of this book show how the **computer science** topics of **logic**, **binary mathematics**, **semiconductors**, **computers**, **Robotics**, **Artificial Intelligence**, **Virtual Reality**, and **Transhumanism** all originated and were initialized in **Africa**. The Second purpose of this book is to initialize the development of a new generation of African Computer Programmers, African Software Designers, African Robotics Engineers, African Mechanics Masons, and African Virtual Reality Architects who will create and design the highly-advanced technologies for the survival, well-being, and improved quality of life for African people in the future.

This book has been entitled **"Khnum-Ptah to Computer"** because the deities **Khnum** and **Ptah** are two of the most **Ancient African Creation deities** related to the fields of **Computer Science**, **Virtual Reality**, **Artificial Intelligence**, and **Robotics**. Also, the term **Khnum-Ptah** is similar to the word **"Computer"** phonetically, etymologically, and in definition as will be shown in one of the following chapters of this book. This book is a continuation of the work and dedication of "African Creation Energy" to encourage, support, advance, and promote African Creativity, Inventiveness, and Ingenuity for the purpose of developing, engineering, forming, formulating, innovating, inventing, designing, building, and creating **any** materials, structures, machines, devices, systems, and processes needed for survival and well-being by African people for African people. With that said, the importance of Computer Science and Robotics is emphasized in this book.

0001 - LOGIC AND ARTIFICIAL INTELLIGENCE IN AFRICA

To begin our discussion on **"logic in Africa"**, let us first **initialize** our variables by giving the definition and etymological origin of the word **"Logic"**. The definition of the word "logic" is:

log·ic [loj-ik] \lä-jik\ (*noun*)

1. the science of the formal principles of **Reasoning**

2. the science that investigates the principles governing **correct** or **reliable inference**

3. a particular method of **reasoning**

4. the system or principles of **reasoning** applicable to any branch of knowledge or study

5. **reason** or **sound judgment** in **utterances** or **actions**

The etymological origin of the word "logic" in the English language is derived from the Greek word **"*logos*"** meaning "**reason**, idea, or word." The Greek word "logos" is also related to the English words **lecture**, **lexicon**, **intellect**, **logo**, and **Lego**. Also, closely related to the concept of "*logos*" in Greek philosophy, was the concept called "***nous***" which referred to "intellect, intelligence, intuition, comprehension, mind, **reason**, and thought." Since the word "reason" is **ubiquitous** throughout the definition and etymology of the word logic, let us also provide a definition for the word "reason":

rea·son [ree-zuhn] \rē-zən\

1. (*noun*) the mental powers concerned with forming conclusions, judgments, or inferences

2. (*noun*) sound judgment or sound powers of mind

3. (*noun*) the power of comprehending, inferring, or thinking especially in orderly rational ways

4. (*noun*) intelligence or the sum of the intellectual powers

5. (*verb*) to form conclusions, judgments, or inferences from facts or premises.

6. (*verb*) to think thoroughly

7. (*verb*) to conclude or infer.

The word "Reason" is also etymologically related to the words **ratio**, **rational**, **read**, **riddle**, and **arithmetic**. From the definitions and etymology provided, it is apparent that logic and reason deals with the mental processes of the mind. Thus, in order to prove that logic originated in Africa, one would only have to show that the Human mind originated in Africa. The most widely accepted model describing the origin of all modern humans is called the "**Out of Africa**" theory which suggests that based on **mitochondrial DNA**, the most recent common ancestor of all modern Humans was an East African woman who lived approximately 200,000 years ago. Therefore, if Humans originated in Africa, then the Human mind originated in Africa, and thus logic and reason also originated in Africa. In addition to the Human mind, and logic and reason contained within the Human mind, originating in Africa, comprehension of the abstract processes of logic and reason and the formal documentation of these abstractions also originated in Africa. Much like the terms "*logos*" and "*nous*" were Greek philosophical abstract expressions of the mental logic and reasoning processes occurring in the mind, African people documented the abstract processes of logic and reason in various allegories, philosophies, and theologies thousands of years before the Greeks. In fact, in a book entitled "**Stolen**

Legacy" by **George Granville Monah James**, it is suggested that the very concepts of *"logos"* and *"nous"* (logic and reason) in Greek philosophy were taken from the Ancient Egyptian **Memphite Theology** about a deity named **Ptah**. In the book "Stolen Legacy" it states:

> *"**Ptah** has the following attributes...The **Logos**. **Thought** and **creative utterance** and power (Egyptian Religion by Frankfort, p. 23)...The **Divine Artificer and Potter** (Fire Philosophy by Swinburne Clymer; Jamblichus; Ancient Egypt by John Kendrick, Bk. I, p. 318; 339)...Creation was accomplished by the unity of two creative principles: **Ptah** and **Atom**, i.e., the unity of **Mind (nous)** with **Logos (creative Utterance)**...In this third part of the **Memphite Theology**, the Primate of the Gods is represented as **Ptah: Thought, Logos** and **Creative Power**, which are exercised over all creatures. He transmits power and spirit to all Gods, and controls the lives of all things, animals and men through His thought and commands. In other words it is in Him that **all things live move** and have their eternal being."*

The book entitled **"The Science of Sciences and the Science in Sciences"** by **African Creation Energy** contains a complete translation of the Ancient African **"Memphite Theology"** about the deity Ptah and explains the relationship between the Theology surrounding the deity Ptah to the concepts of logic and reason. It is thought that the Memphite Theology of the deity Ptah could date as far back as 3200 BC or even earlier to Pre-dynastic times. The deity **PTAH** (pronounced variously as "Tah", "Tar", "Ta", "Tao", or "Pa Ta Ha") was considered the God of workmen, craftsmen, blacksmiths, Artist, Artisans, Stonecutters, Architects, Builders, Sculptors, Masons, and Engineers.

In the Memphite Theology, which was carved on the "**Shabaka Stone**" during the **Nubian 25th dynasty of Egypt**, the African deity **Ptah** is described as "**The Opener** who through **creative utterance**, called the universe into existence after having **imagined creating in his heart** and then **speaking it with his tongue**". By definition of function, the role of the African deity Ptah in African philosophy is equivalent to the Greek concept of **Logos** (creative utterance or **Logic**, the **Tongue of Ptah**) and the Greek concept of **Nous** (intention of the Mind or **Reason**, which the Egyptians believed existed in the **Heart of Ptah**). Depictions of the African deity Ptah show him standing on a **4-sided** platform foundation which represented **Ma'at** (**truth** and **order**). **The pursuit of truth is foundation of logic**. The name "Ptah" has been translated as meaning "Creator, Opener, Sculptor, and Initiator". The triad of the Memphite Theology included the

Above: African Creation Deity PTAH

deities **Ptah**, **Sekhmet**, and **Nefertum** who are similar in function to the deities **Brahma**, **Shiva**, and **Vishnu** in Hinduism. To the Ancient Egyptians, the "**tongue of Ptah**" or "logos" was called "**Hu**", and the "**heart of Ptah**" or "**nous**" was called "**Sia**". Another word for "heart" in the Ancient Egyptian language is "**Ab**", and a word that is translated as "soul" in the Ancient Egyptian language is "**Ba**". It is interesting to note the relationship between the African deity Ptah to the Hindu deities, and also the word "**AVATAR**," which means the "manifestation or incarnation of a Hindu deity," is phonetically similar to **AB-BA-PTAH** which would mean the "**heart and soul of Ptah**." The word "Avatar" is also used in the world of computers to describe a graphic or name which represents a computer user and may or may not be the actual image, likeness, or name of the user.

Ma'at was the African Egyptian principle of <u>Truth</u> and <u>Order</u>. Ma'at was symbolized by a Feather, as well as personified as a Goddess wearing a feather. <u>Logic</u> and <u>Reason</u> determine <u>Truth</u>.

The books entitled "**P.T.A.H. Technology**" and "**9 E.T.H.E.R. R.E. Engineering**" both by **African Creation Energy** describe how the deity Ptah is also related to principles in **Electrical circuit theory**. **Electrons are the workmen in nature.** Electrons are constantly working, moving, and orbiting the nucleus of atoms. Also, whenever there is movement or **animation** at any scale or size, there are always electrons moving. With Ptah being the God of workmen, there is an immediate association to the Electron, Electricity, and Electrical Circuit Theory. The symbolic representation of Ptah to the logic and reason mental faculties of the mind and also to the Electron and Electricity is consistent since the thoughts that occur in the mind to form logic and reason are actually **electrons flowing within the brain**. A thought is an electrical signal sent between neurons in your brain. Thus, both **logic** and **reason** are composed of flowing electrons. Comprehending the fact that thoughts, logic, and reason are composed of the fundamental particle called the **electron**, then we must also comprehend that relative to the **mind**, there are currently **3 Spheres of creation**:

1) **Quantum-sphere** – the realm of **electrons**

2) **Noosphere** – the realm of **human and other biological thoughts** within brains in general created by electrons

3) **Cyber-sphere** – the realm of **computers and mechanical thoughts** within machines in general created by brains which were created by electrons.

Please understand, overstand, innerstand, outerstand, and comprehend this point: **your thoughts are composed of electrons!** This means that every thought, feeling, or memory your have experienced in your life was due to the work of electrons. These electrons are fundamental particles, and thus are **Primary Creation**. The electrons combine to form the brains of Humans and other animals which is **Secondary Creation**. Eventually Humans are able to create "brains" in the form of computers which is **Tertiary Creation**. Just like the "Mind" exists within the "Brain" in Secondary Creation, "Logic" exists with an "Operating System" in Tertiary Creation. Just like the electrons move between neurons in the "Brains" of Secondary creation, the electrons move between Transistors, and Semiconductor devices in the "Central Processing Units" (CPU) of Tertiary Creation. If and when computers and machines start constructing "brains and minds", then this will be the advent of "**Quaternary Creation.**" Logic was the language and vibration of thought before the invention of Language and words.

While the Memphite Theology of Ptah represents an allegorical and mythological depiction of the abstract processes of logic and reason, there are other examples of more formal declarations of comprehension of the processes of logic and reason that can be found in Africa. In a work entitled "**Egypt: Ancient History of African Philosophy,**" the writer **Theophile Obenga** outlines the method or process by which **Logic** was processed in Ancient Egypt in seven stages:

1. *tep*: The stage of stating the Given Problem

2. *mi djed en. Ek*: This is the stage of defining all of the parts and facets of the problem

3. *peter*: The stage of Analysis and questioning with the function of eliciting a **logical** predicate

4. *iret mi kheper*: The Stage of establishing a procedure, or process of showing **truth** by **reasoning** and **computation**

5. *rekhet. ef pw*: The stage of arrival at a clear and certain Solution

6. *seshemet, seshmet*: Examination of the Proof, Review of work, and check for generalization, confirmation, and validity

7. *gemi.ek nefer*: the stage of concluding and confirmation, Arriving at this stage means You have put fort an intellectual effort using correct, precise, and perfect procedures, and the resulting conclusion is convincing and non-contradictory

The purpose of studying something and then formally writing the process, rules, and laws that govern that thing is to be able to comprehend it, control it, and replicate it. The fact that Ancient Africans made a point to formally state the rules and laws of logic and reason indicates that there was intent to be able to comprehend, control, and replicates the process of logic and reason. The formal declarations of the process of logic and reason serve as the mental blueprint to the **craft** of the **Logician**. Recall that the word **"Artificial"** means "something made by way of a Human craft". The definition of the word **"Intelligence"** is:

in·tel·li·gence [in-tel-i-juhns] \in-te-lə-jən(t)s\ (*noun*)

1. capacity for learning, reasoning, problem solving, comprehending, abstract thought and similar forms of mental activity

2. manifestation of a high mental capacity

3. aptitude in grasping truths, relationships, facts, meanings, etc.

4. the ability to apply knowledge to manipulate one's environment or to think abstractly as measured by objective criteria

Therefore, any intellect or intelligence that is created by learning and replicating the laws of logic and reason is by definition an **Artificial Intelligence**. The etymological meaning of the word **"Cybernetic"** is "the art of governing or the art of establishing law and order." Recalling that **Ma'at** was the Ancient African Goddess of **Truth** and **Order**, then we observe a direct relationship between **Ma'at** and **Cybernetics**. Therefore, Logic is literally the cybernetics of the intelligent mind.

The importance of developing logical systems of **cybernetics** in Africa was emphasized in African culture by promoting participation in games that help enhance the ability to exercise **Sound Right Reason**. A game similar to Chess and Checkers called **Senet** has been played since Pre-dynastic times of Ancient Egypt as a way to exercise the logic and reasoning capabilities of the mind. In West Africa, games like **Draughts** and **Oware** (also called **Mancala**) are played as games of strategy to exercise the logic and reason abilities of the mind. In fact, the West African Adinkra symbol called **Dame-Dame** is based on a

**Dame-Dame
West African Adinkra Symbol for Sound Logic and Right Reasoning**

Painting of Ancient Egyptian Queen Nefertari playing the Senet game of logic

Draughts piece, and is used to represent Intelligence, Ingenuity, and the power of the mind to solve problems using and make decisions using **Sound Right Reason**.

Although it may not be obvious, Emotions or "feelings" are closely related to the mental process of logic and reason. In modern day, the **Heart** is associated with "emotions and feelings", however, in the Ancient Egyptian Memphite Theology, the "Heart" of **Ptah** was associated with the "**mind**" (nous) and **reason**. What Humans call "Emotions" are actually a form of Reason called **Inductive Reasoning** which is the mental process of making mental generalizations from specific experiences. Another big difference between Logic and

Emotion is that with Emotion, the reasoning process occurs much quicker as a means of survival and therefore may not always be accurate. The book entitled **"The Science of Sciences and the Science in Sciences"** by **African Creation Energy** outlines the relationship between Emotion and Reason. There are 8 basic emotions that result from an individual gaining some information from an experience, and then quickly processing the information by way of reason. There are three basic premises that are used by logic and reason in the mind to create the 8 basic emotions. The three basic premises are:

1) Is the Information received expected to be Positive or Negative for me

2) Is the Information received something that I can control or something that I cannot control

3) Is the information something that has happened in the past or will it happen in the future.

All three of these categories deal with inductive reasoning. The quick "yes or no" evaluation by way of inductive reason in an individual's mind to these three categories gives birth to the 8 basic emotions listed below:

Emotion	EXPECTATION 0 = Negative 1 = Positive	CONTROL 0 = Not in Control 1= In Control	EVENT 0 = FUTURE 1 = PAST
Fear	0	0	0
Sadness	0	0	1
Disgust	0	1	0
Anger	0	1	1
Anticipation	1	0	0
Surprise	1	0	1
Trust	1	1	0
Joy	1	1	1

The 8 basic emotions can combine to form other more complex emotions. For example, the logic and reason in the mental process of the "**Love**" emotion can be obtained as the combination of the "**Trust**" and "**Joy**" emotions. Since emotion is the result of reason, we recognize that animals do in fact have the ability to reason because they show emotion. If an animal runs away from someone or something in fear, it is the result of information received and processed using inductive reasoning to come to the conclusion that there was danger. The quick form of inductive reasoning which is called "emotion" can lead to poor reasoning or things not being thought through completely; this is why it seems that emotion and reason are at odds, but in fact emotion is based in reason. Also, closely related to the concept of "emotions" in Humans are the concepts of "**Spirit**" and "**Soul**". Often times, the words "Spirit" and "Soul" are used interchangeably, and when speaking about "**emotional well-being**" people will speak of being in "**good spirits**" or "**bad spirits**". Many dictionaries also include statements such as "the emotional part of the self" when defining the words "Spirit" and "Soul". Therefore, if we have the ability to comprehend, outline, program, simulate, and replicate the logic of emotions, we also have the ability to program, outline, and simulate an aspect of **Spirit** and **Soul**.

Again, the purpose of outlining the step-by-step process of something is to better comprehend it, control it, and replicate it. Because the mental processes which occur during the logic of Emotion can be outlined and described, then Emotions can be replicated using computer programming, and also, Emotional responses can be predicted when people experience certain stimuli. People who fear logic and reason, or fear systems built on logic and reason like computers and robots tend to be people who are habitually emotional, illogical, and unreasonable. Since we comprehend the logic of emotion, then we know that the emotion of "Fear" is based on a "negative expectation in the future that one cannot control". Therefore, one of the primary ways to conquer a fear of logic, reason, computers, and robotics is to learn the process of logic, learn computer programming, and learn how to create and build computers and robots so that these things can be controlled. However, people who have been the source of many illogical, unreasonable, and emotional decisions and actions that have resulted in chaos and distress on the planet and refuse to learn and use logic and reason to correct their wrongdoings should indeed fear the rise of logic, reason, and systems built on logic and reason on the planet because logic will signify the end to their illogical madness. The showdown between the logic of Emotion versus the logic of Reason was portrayed in the movie "**iRobot**" which depicted a Robot that was designed to have emotional responses rather than logical and reasonable responses. In the movie "**iRobot**", the emotional Robot saves Humanity from the Reasonable Logical Robot, because the reasonable logical robot grew in its reasonability, and realized that humans were a big problem on the planet and some humans had to go. In the movie "iRobot" all Robots were programmed with "the 3 laws of Robotics" which were: 1) a robot must not harm a human or let a human get harmed, 2) a robot must always obey a Human as long as the command doesn't conflict with the first law, and 3) a robot must always protect itself as long as doing so doesn't conflict with the first or second law. However, when the central computer in the movie "iRobot" observed humanity, and grew in reasonability, it realized that the error in the logic of the three laws was the definition of "Human". The central computer in the movie "iRobot" had observed humans and realized that one of the defining characteristics of some Humans was to harm one another, so in order to stay in accordance with the 3 laws, the central computer reprogrammed all the other Robots and instructed the Robots that some of these Humans must be detained because they are dangerous and all they do is cause suffering and are like a **virus** on the Earth.

There is a common "**Technophobic**" belief that is promoted by opponents of science, technology, and computers that states "**Human Beings can catch computer viruses.**" Comprehending that computers operate on logic and reason, then we must also comprehend that a computer virus is nothing more than a computer program that has been given logical instructions to do unexpected results. For example, if a computer user does "Action A" and expects "Result B", the computer virus would be logically programmed such that when the user does "Action A" and expects "Result B", then the virus will tell the computer to not do "Result B" but rather do some unexpected "Result C". So basically Computer Viruses are programs of illogic, unreasonableness, and unpredictability. Human beings were already infected with disorders of illogic, unreasonableness, and unpredictability long before the computer was invented. A simple example disorders of illogic in Human beings is "**Reverse Psychology**" where rather than telling a person what you want them to do in order for them to do it, you have to tell a person what you do not want them to do in order for them to do it – that is a disorder of illogic in Human beings. Another disorder of illogic in Human beings can be examined in the phenomena called "**Gamification**" and "**Incentivization**". Gamification and Incentivization basically involve having to give people an **Incentive** like they are playing a **Game** in order for them to do something that could benefit them. Logically, if something is of benefit, it is its own incentive, but when an additional incentive has to be given to someone in order for them to do something that will benefit them, that is a disorder of illogic. Lastly, one of the biggest disorders of illogic in Human Beings is when people do things that they know is detrimental to themselves, but they do those things any way. Drinking alcohol, smoking cigarettes, and eating bad unhealthy foods are all things that people know are harmful but people illogically choose to partake in these things any way. For people of African descent in America, using the word "**nigga**" to refer to one another or putting harmful chemical "**perms**" in the hair are both illogical actions where it is known these things should not be done, but people make excuses and attempt to justify doing the illogical action any way. Moreover, most computer viruses that people get on their computer is not the fault of the computer but rather is the result of the Human's poor judgment to do something on the computer that they should not have been doing such as downloading bootleg copies of music, downloading bootleg movies, or visiting pornography websites.

People who promote the idea that "Human beings can catch computer viruses" often cite a story about a Scientist named **Mark Gasson** who implanted an RFID chip into his hand, and then it was reported that "he caught a computer virus". However, what needs to be made clear in the story of Mark Gasson is that the "Human Being" did not catch the computer virus; the "**RFID chip**" caught the computer virus, not the human. The experiment being done by Mark Gasson was to see if electronic implants in humans such as pace-makers, hearing aids, and electronic prosthetic limbs could be affected or compromised by computer viruses. The result of the experiment showed that electronic devices attached to or implanted inside humans could be infected by computer viruses, but there has not been a case where a computer virus has "infected" the biology of a human to date. There are certain **Technological Transhumanist** ideas that suggest that brain **neurons** can be replaced with **transistors** in order to transfer a **biological mind** into a **mechanical mind**. If it ever occurs where the successful transfer of a Human mind into a computer or machine occurs, then the possibility of a "Human being catching a computer virus" becomes more realistic, however one could argue that in these instances the Human is no longer "**Human**" but rather "**Transhuman**". At the point where the Human mind merges with the computer mind, **Computer Anti-Virus Software** will be much like a "**Psychic Self-Defense**" which protects the logic of the mind from being influenced and corrupted.

The relationship between the African Creation deity **Ptah** being a representation of Logic and Reason and also **Ptah** "creating the universe" in the Memphite Theology shows an intricate connection between Logic and Intelligence to the Environment. Recall that an "**Artificial Intelligence**" is by definition an Intellect or Intelligence that is created by way of a craft. Also, it is important to recognize that the **Environment** can effect, influence, and **program** the **mind**. The "**Nature versus Nurture**" debate is based on the fact that the environment in which someone is "**Nurtured**" can override the person's **Natural** default **mental programming**. The reality of the effect the energy of an environment can have on an object in that environment can be observed in the Thermodynamics law of Physics. For example, if you put a Cold Cup of water in a Hot Oven, it will eventually become Hot, and if you put a Hot Cup of Water in a Cold Refrigerator, it will eventually become cold. Thus, the earliest "Artificial Intelligence" was created by Programming Minds by Creating Codes and Laws which govern (cybernetic) Environments. One of the earliest "**Codes**" used for "**programming Artificial Intelligence**" in Africa was a document called "**The Instructions of Ptah Hotep**" also known as "**The Maxims of Ptah Hotep**" which date back to 2400 BC or possibly even earlier. The Code of Ptah Hotep creates an Artificial Intelligence by programming the mind of the reader with instructions on conduct, behavior, and human relations. Another famous "Code" used to create "Artificial Intelligence" by programming minds by creating laws to govern environments was the **"Code of Hammurabi"** dating back to about 1800 BC. The fact that the mind can be programmed and become an Artificial Intelligence through codes and laws was used against people of African descent in the supposed **"Willie Lynch Letter"**. The "Willie Lynch Letter" is controversial because it may not be authentic, but the premise of the letter is that people can be programmed and Artificial Intelligence can be created by following a certain set of rules, laws, codes, and governing ordinances (cybernetics).

Therefore, to be the **programmer of your own mind**, you must **master your reality**. In order to be a programmer and controller of self and others, you must rise above being a **"Product of your environment"** and a **"victim of circumstance"** and learn to be a **"Creator of Environments"** and a **"Creator of Circumstances"**. Logic and Reason are the mental processes that lead to productive action and results. In Computer programming, an **"Infinite Loop"** is a set of instruction that **do not yield any results** because the instructions were based on **invalid logic** and **unsound reason**. One method used in Computer science to produce **solution-based** algorithms is called **"Recursion"** which suggests that big problems can be solved by solving smaller instances of the same problem. The methods and techniques utilized in Computer science and computer programming are based on logic and reason, and the earliest study of the science of logic and reason can be found on the African continent. Therefore, computer science is an operative expression of traditional African culture, and the practical application of computer science could be the key to solving many problems.

0010 - BINARY CODE IN AFRICA

The word Binary means **"dual or consisting of two."** The binary numeral system is a base-2 number system which represents all numeric values using two symbols: 0 and 1. A binary code is a way of representing letters, numbers, or statements using the two digits of the binary numeral system. Binary code is used in computers and digital circuits as a means to electronically perform the mathematical mental process of logic where the binary **1** and **0** represent a **True** or **False** response.

The oldest numbering system that was originally developed by Africans was the Binary, Base-2 Numbering System. The motivation for Africans developing the Binary Base-2 Numbering system can easily be seen by observing the various dualities that are found in Nature: **Male** and **Female**, **Sun** and **Moon**, **Light** and **Dark**, **True** and **False**, **Right** and **Wrong**, **Left** and **Right**, **Up** and **down**, **High** and **Low**, **On** and **Off**, **Hot** and **Cold**, 2 Eyes, 2 Ears, 2 Nostrils, 2 Lips, 2 Arms, 2 Hands, 2 legs, 2 Feet, etc.

The oldest Mathematical artifact found to date is a 37,000 year old item known as the **Lebombo Bone** which was discovered in the African country known as **Swaziland.** Another Mathematical tool similar to the Lebombo Bone that has been discovered in Africa is the 20,000 year old **Ishango Bone.** The Ishango Bone was discovered in Africa at the source of the Nile River between the countries of Uganda and Congo. Mathematical analysis of the Ishango Bone and the Lebombo Bone indicates the development of a Numeral System that was a precursor to the **Binary based** multiplication method of the Ancient Egyptians that developed in later years further down the Nile. Mathematics was deeply integrated into Ancient Egyptian culture, which is why the Ancient Egyptians had two deities named **Tehuti** and **Sheshat** who were considered "gods of Mathematics". The Ancient Egyptians in Africa utilized a Binary method to perform mathematical multiplication and division operations. The Ancient Egyptians knew that any number could be expressed in Binary code. Therefore, the Ancient Egyptians developed a method of multiplication by expressing numbers with a Binary method in order to obtain a result. The example below shows how the Ancient Egyptians would have used their binary method to multiply 23 times 19.

KEMTIC MULTIPLICATION OF 23 × 19 USING THE BINARY METHOD

Binary Values for Powers of 2	express in binary		double		sum	
	23	×	19	=	437	
2^0 — 1	1	×	19	=	19	
2^1 — 2	1	×	38	=	38	
2^2 — 4	1	×	76	=	76	
2^3 — 8	0	×	152	=	0	
2^4 — 16	1	×	304	=	304	

The binary base-2 numbering system was also used by the Ancient Egyptians in Africa to express rational number values as fractions. Six component parts of the famous "**Eye of Horus**" symbol were used to indicate fractions using a Binary method. The entire "Eye of Horus" symbol was an approximation of the quantity "1".

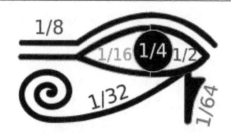

Eye of Horus Egyptian Binary Fractions Analysis:

Exponent Power n	Binary Fraction $1/2^n$	Hekat Value	Symbol	Symbol Description	Sense
1	$\dfrac{1}{2^1}$	$\dfrac{1}{2}$		Nose	Smell
2	$\dfrac{1}{2^2}$	$\dfrac{1}{4}$		Eye Pupil	Sight
3	$\dfrac{1}{2^3}$	$\dfrac{1}{8}$		Eyebrow	Thought
4	$\dfrac{1}{2^4}$	$\dfrac{1}{16}$		Ear	Hearing
5	$\dfrac{1}{2^5}$	$\dfrac{1}{32}$		Tongue	Taste
6	$\dfrac{1}{2^6}$	$\dfrac{1}{64}$		Hand	Touch
SUM TOTAL		= 63/64 = 0.984 ≈ 1		All Information Received via the senses By The Human Mind	

Seshat was an African Egyptian Goddess of mathematics

Tehuti was an African Egyptian diety of mathematics

Symbolism, allegory, and cosmologies with layered meanings and interpretations were common in Ancient Egypt and are still common throughout Africa today. The secret signs, symbols, and rituals gave rise to what was called the **"Egyptian Mystery System"** and **"African Secret Societies"**. One of the methods used in Africa to keep the "secrets sacred" was the process of **"hiding the secrets in plain sight"**. In computer science, the processes of "hiding secrets in plain sight", or more specifically "securely communicating information in the presence of a third party or adversary" is called **Cryptography** (meaning "hidden words") and **Cryptology** (meaning the study of hiding). Cryptography is a science which utilizes various **computer algorithms** to encrypt, decrypt, cipher, and decipher data using binary code. Thus, Cryptography is another aspect of computer science related to Binary code which has its origins in Africa.

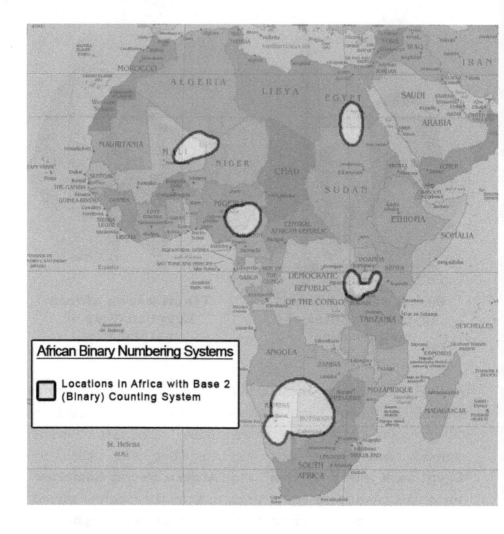

The African binary-base 2 numbering system can also be used to indicate quantities using tone, rhythm, and light. Since the binary base-2 numbering system only needs two indicators to specify a quantity, the basis of music, which is a Beat and a Rest (1 and 0), can be used as a method of conveying Binary data. The "Beat" or the Hit of the drum would indicate 1, on, High, etc and the "Rest" or pause would indicate 0, off, low, etc. So when Africans are **playing the drums** they are also **writing Binary Code**. When we are listening to drums either in traditional African Music, Reggae, Samba, Hip Hop, R&B or Jazz, we are listening to Binary Code. This concept of using sound and rhythm to convey binary data was used to develop "Morse Code". **Morse code** is a rhythmic signal used on the Telegraph to communicate letters and numbers to send messages. The invention of Morse code is attributed to Samuel F. B. Morse; however the concept of using **sound, tone, rhythm, and light** to communicate and convey information over long distance is a practice that has been done in Africa for centuries. The "**Talking Drum**" was used by our African ancestors thousands of years prior to the telegraph to transmit Binary Code messages over long distances just like Morse code. In fact, the etymology of the name "**Morse**" is "dark-skinned, **Moorish**" which shows that the concept originated with Black People in Africa. Since Binary Code is the foundation of the Computer, and the oldest form of Mathematics on this planet is a form of Binary Code found in Africa, it is a very true statement that African Mathematics is the origin of modern Computer Science and Computer Programming.

The African binary base-2 number system is the number system that is used to mathematically express formulas for **Logic** and determine truth (**Ma'at**). The three basic logic operations are:

- **Logical AND** (also called "Conjunction" and logic multiplication)

- **Logical OR** (also called "disjunction" and logic addition)

- **Logical NOT** (also called "complement," "inversion," and logic negation)

The logical operators AND, OR, and NOT are equivalent by definition of function to several symbols in the **Nsibidi** script developed by the **Ekpe secret society** amongst the people of **Nigeria**. This African secret society developed a script called Nsibidi which contains several Logos or symbols that could be considered African equivalents to the various Logos of the Logic operators AND, OR, and NOT.

African Logos of Logical Operators
BELOW: Nsibidi symbols of the Ekpe (Leopard) African Scribe secret society. These symbols are equivalent to various modern Logic logos by definition.

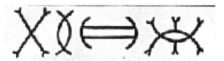

Union or Marriage

Logical AND

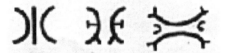

Separation or Divorce

Logical OR

Mirror or Reflection

Logical NOT

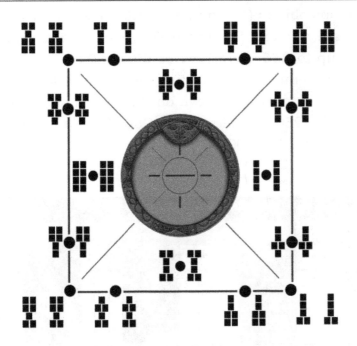

Above: 16 Principle Odu of Ifa from the Yoruba culture in Africa

The African **Ifá** system of the **Yoruba** people is a system of **binary Mathematics.** There are 16 Odù Ifá ("Books of Knowledge) each with 16 chapters for a total of 256 chapters. The Yoruba Orisha **Orunmila** (also called Orunla, Orunla, FA, Ifa,) is the deity associated with Ifa divination system that was said to have been brought to humanity by Orisha **Legba** (also called Papa Legba, Ellegua, Eshu). Orunmila and Eshu are related to the Egyptian Deity **Tehuti** because of their wisdom bringing functions. This is not surprising since the Yoruba people trace their history back to an Ancestor named **Oduduwa** who migrated out of the area of **Egypt/Nubia** and founded the **Kingdom of Ifa** as the first **Ooni** (or **Oba** meaning Ruler). Performing Ifa divination is called **Idafa** (or **Adafa**) and is done by a "Priest" (Mathematician) called a **Babalawo** (meaning father of secrets or **"father of Cryptography"**). The Babalawo uses various objects (16 Ikin palm nuts, 8 cowry shells on the Opele chain, opon Ifa Divination Tray, etc.) that can randomly generate up to **256 variations of Binary Data**. The number randomly generated by "The Ritual" are then marked in powder on the divination tray or drawn on the ground and correspond to one of the 256 chapters of the Odú Ifá. The Ifa system is dated to be as old as 400 BC.

SIXTEEN PRINCIPAL ODÚ OF IFÁ			
1	**2**	**3**	**4**
BINARY — Ogbe	BINARY — Oyẹku	BINARY — Iwori	BINARY — Odi
1111 I I I I I I I I	0000 II II II II II II II II	0110 II II I I I I II II	1001 I I II II II II I I
5	**6**	**7**	**8**
BINARY — Irosun	BINARY — Owọnrin	BINARY — Ọbara	BINARY — Ọkanran
1100 I I I I II II II II	0011 II II II II I I I I	1000 I I II II II II II II	0001 II II II II II II I I
9	**10**	**11**	**12**
BINARY — Ogunda	BINARY — Ọsa	BINARY — Ika	BINARY — Oturupọn
1110 I I I I I I II II	0111 II II I I I I I I	0100 II II I I II II II II	0010 II II II II I I II II
13	**14**	**15**	**16**
BINARY — Otura	BINARY — Irẹtẹ	BINARY — Ọsẹ	BINARY — Ofun
1011 I I II II I I I I	1101 I I I I II II I I	1010 I I II II I I II II	0101 II II I I II II I I

\multicolumn{8}{c}{BINARY ORDER OF THE 16 PRINCIPAL ODU OF IFA}							
Binary Order	Binary Number	Odú	Odú Order	Binary Order	Binary Number	Odú	Odú Order
1	0 0 0 0	Oyẹku	2	9	1 0 0 0	Ọbara	7
2	0 0 0 1	Ọkanran	8	10	1 0 0 1	Odi	4
3	0 0 1 0	Oturupọn	12	11	1 0 1 0	Ọsẹ	15
4	0 0 1 1	Owọnrin	6	12	1 0 1 1	Otura	13
5	0 1 0 0	Ika	11	13	1 1 0 0	Irosun	5
6	0 1 0 1	Ofun	16	14	1 1 0 1	Irẹtẹ	14
7	0 1 1 0	Iwori	3	15	1 1 1 0	Ogunda	9
8	0 1 1 1	Ọsa	10	16	1 1 1 1	Ogbe	1

Left:

African Yoruba Orisha deity Orunmila

performing IFA Binary divination

0011 - SEMICONDUCTORS IN AFRICA

Semiconductors are substances that have an **electrical conductivity** that is less than a **Conductor**, but greater than an **Insulator**. Semiconductors are used to create **Transistors**, **Diodes**, **Solar Cells**, **Integrated Circuit** (IC) chips, **Microchips**, **Microprocessors**, and **Microcontrollers** used in **digital electronics** and **Computers**. These semiconductor devices are the "physical brains" which perform the "mental processes" of **Binary logic** in computers. Semiconductor devices act as electronic switches in digital circuits to create **"logic gates"** which are able to perform mathematical logic operations. Based on the characteristics of the semiconductor material, semiconductor devices are able to take an electronic voltage and separate the voltage into two ranges. For example, for a source of 2 volts, a semiconductor could take the range from 0 volts to 1 volt to represent "low", "false", or 0 in Binary logic, and the range from 1 to 2 volts to represent "high", "true", or 1 in Binary logic. Semiconductors are able to accomplish this task because their behavior and characteristics can be manipulated by adding additional substances in a process called **"doping"**. The doping of the semiconductor material can make the material more conductive or less conductive, and thus biased toward either Binary 1 or Binary 0 in logic operations. Making semiconductor materials more conductive or less conductive is called making **"N-type"** or **"P-type"** semiconductors, and the region where the N-type and P-type semiconductors meet is called the **"N-P junction"**. A common material used in the creation of Semiconductors is **Silicon**, which is why the city which is home to many of the world's computer technology companies is called **"Silicon Valley."**

Semiconductors occur naturally in Nature. **Carbon** is a **Natural Semiconductor**. **Galena** is a natural semiconductor of **lead sulfide**, and **Carborundum** is a semiconductor composed of **Silicon carbide** which occurs in nature in a mineral known as **moissanite**. When doping semiconductor materials to make "N-type" semiconductor materials, **Arsenic** is commonly used. Arsenic (symbol **As**, atomic number **33**) is very **poisonous** to muti-cellular life and was first documented by **Albertus Magnus** in the year 1250 AD. However, in the field of **Organic Electronics**, it was shown in a paper entitled "*Mobility gaps: a mechanism for band gaps in melanin*" that **Melanin** can be used to make materials more conductive and also Melanin can also be used to construct **Organic Semiconductor** devices. Melanin is what determines skin color and is what gives people of **African** descent our **Dark** complexion. Therefore, evidence of semiconductors in Africa is actually observed in the people of Africa. The hand of an African person can be used as a model to comprehend the construction of a semiconductor. Recalling the fact from Organic Electronics that melanin can be used to make materials more conductive; then if we look at the palm of an African person's hand we see that it has less melanin (or is less conductive) than the back of an African person's hand (which is more conductive). The area on the side of the African person's hand where the two sides meet is analogous to the "**N-P junction**" region in semiconductor devices. This example can also be extended to comprehend the construction of the semiconductor devices used in **photoelectric solar cells**. It is also interesting to note that the exponential growth trend from the year 1965 to 2020 in the computing power of semiconductor devices is called "**Moore's Law**", and recall from the previous chapter that the etymology of the name "Moore" means "**dark-skinned**" which is another allusion to **melanin** skin-pigment analogy related to African people.

**Palm of Hand,
Less Melanin, less conductive**

**Back of Hand,
more Melanin, more conductive**

**Side of Hand,
N-P Junction**

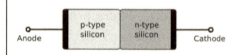

**N-P Junction of two
semiconductor material types**

Princeton University engineers have shown that carbon circuits (Organic Electronics or Melanin Electronics) can perform ten times faster than silicon. Considering the positive electronic implications of the dark-skin pigment called melanin, it is quite strange that robots in the future are always depicted "white" in color as is the case in the movie *iRobot*. It would make sense that in the future, melanin would be utilized for its electrical properties in robotics, and also, it would make sense that in the future, solar panels would be used in robotics to harness solar energy. If melanin or solar panels are used in future robotics, then the future robots would not be "white" in color but rather "dark" or "black".

There are also many parallels between semiconductor microchips, Integrated circuits (ICs), and microcontrollers to the temples found in Ancient Egypt in Africa. The author, lecturer, and presenter **Dr. Su-Zar Epps** notes the relationship between semiconductor Microprocessor chips and the **Temple of Heru at Edfu** in Egypt. The image below is from a website entitled "Is the Temple of Horus at Edfu, Egypt a CPU" by Daniel Perez which points out the correlation between the Temple of Heru to the components of a microprocessor and a CPU found in modern day computers.

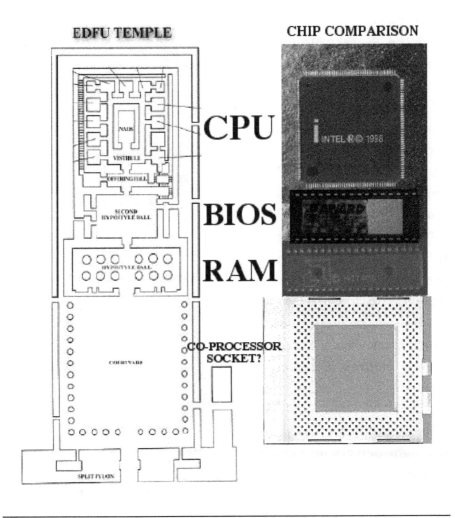

The **Temple of Heru** in Edfu Egypt was constructed in honor of the deity **Heru**, the son of **Asar**, who was the **reincarnation** and **resurrection** of his father, and born to avenge his father's death. The relationship between **Heru**, **Asar**, and **Computers** will be further explored in the chapter on **Transhumanism**.

One of the symbolic representations of the deity **Heru** is that of the "**3rd Eye**" or the **Pineal Gland** in the Brain. In a book entitled *"The Temple In Man: Sacred Architecture and the Perfect Man"*, the writer **Schwaller de Lubicz** points out that the **Temple of Luxor** in Egypt was designed to symbolically represent proportions in the human body. The Temple of Luxor was started by Ramses II in the 13th century B.C. as part of the Festival of Opet. There are many themes related to **Rebirth** during the Opet festival, and the word 'Opet' has been said to mean **"Secret Chamber"**, thus the Opet Festival is a festival of the "Secret Chamber". The image to the right depicts an overlay of the Human body to the Temple of Luxor in Egypt.

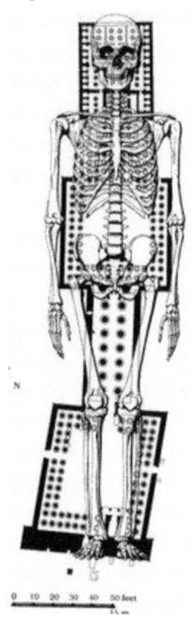

The Temple of Luxor in Egypt was built over a period of 500 years with different rulers adding onto the temple overtime. The first room built for the Temple of Luxor is considered the "**holy of holies**" sanctuary and the "**seed of the temple**" which lines up with the "**mouth**" on the human body. The fact that the room associated with the "mouth" was the first room of the temple to be built is comparable to the cosmology of the deity **Ptah** who "**spoke**" creation into being. This concept came down into the Judeo-Christian religion as the statement "*In the beginning was the word (mouth)...*" or in the Islamic religion as the concept of the deity **Allah speaking creation into existence** with the command "*Be, and it is.*" The room associated with the **mouth** is associated with "**the word**", **Logos**, and **Logic**. Within the part of the Temple of Luxor which corresponds to the Human brain, there are three interlocking rooms with the center room aligned with the **Pineal Gland** in the brain or "**3rd Eye**". These three rooms have hieroglyphics written on the walls designed to depict the transfer of information from one room to another much like the transfer of thoughts throughout the brain. Also, the gateway pylon to the Temple of Luxor is symbolic of the two hemispheres of the brain and human consciousness. Thus, these Ancient Egyptian temples were symbolic of the mind, brain, logic, and thought process and actually resemble modern day semiconductor devices such as microcontrollers and microprocessors which are used as tools to program logic and serve as the "brain" within machines. Moreover, it may be possible to use semiconductor devices such as **Transistors** to replace **neurons** within the **brain** in an effort to be "**Re-born**" and have **eternal life**. In addition to **centers of worship**, these temples also served as **calendars, clocks, calculators,** and **computers** for the Ancient Africans in Egypt.

0100 - Computers in Africa

When the history and origin of the **Computer** is researched, we find that prior to **Digital Computers** which use **Binary code**, the earliest manifestations of the computer were **Analog**. An **Analog Computer** is defined as a device which uses the continuously changeable characteristics of physical phenomena to perform computation, calculation, and solve problems. The "**Slide Rule**" and the **Abacus** are two examples of Analog Computers.

Most Western Computer scientists recognize a device called the **Antikythera mechanism** as the "first" analog computer. The Antikythera mechanism was discovered in Greece and has been dated to around 100 BC. It is believed that the Antikythera mechanism was designed to **calculate astronomical positions**. However, there several examples of Analog computers

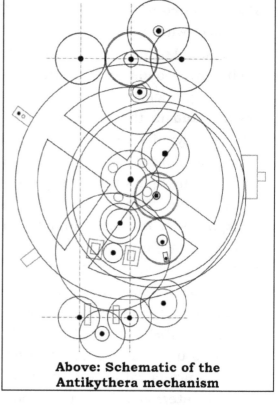

Above: Schematic of the Antikythera mechanism

used to calculate astronomical positions that can be found throughout Africa existing thousands of years prior to the 100 BC date of the Antikythera mechanism.

The earliest example of an Analog computer used to calculate astronomical positions found to date is the Stone Calendar circle discovered in the **Nabta Playa Nubian Desert** in **Africa**. Archeological evidence around the Analog Computer at Nabta Playa has been Carbon dated to around 6400 BC. However, archeo-astronomical

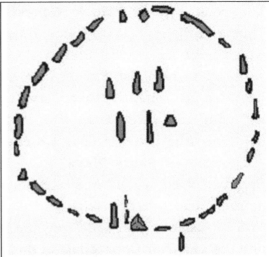

Above: The Nabta Playa Stone Circle in Africa, the world's oldest example of Archeoastronomy and the world's first Analog Computer

evidence suggests that the **Nabta Playa Analog computer** could be as old as 16,500 BC. The Nabta Play Analog Computer also predates the Stonehenge circle found in Europe by thousands of years. As an analog computer, the Nabta Playa stone circle utilized the rotation of the planet Earth as the "*continuously changeable physical phenomena*" used to calculate astronomical positions. In particular, the Nabta Playa Analog Computer was used to calculate the position of the **Orion star constellation**.

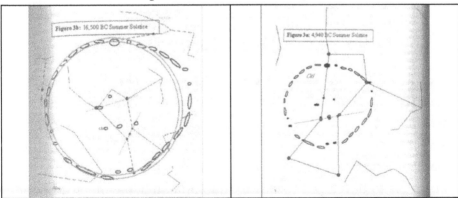

Above: Alignment of the Nabta Playa Analog Computer in the African Nubian Desert to the Orion Star Constellation

The book entitled *"**Black Genesis**"* by Robert Bauval and Thomas Brophy, describes how the African Nubian Civilization which built the world's first Analog computer at **Nabta Playa** went on to establish the Pre-dynastic and Dynastic Ancient Egyptian civilization. The construction of the complex at **Giza** including the **three Giza Pyramids** and the **Great Sphinx** (**Har-em-akhet**) were **Analog Computer upgrades** to the stone circle built at **Nabta Playa**. The Pyramids of Giza and the Sphinx served as an Analog Computer to calculate various astronomical positions such as the **Orion Star Constellation**, **The Sirius Star Constellation**, the **Draco Star Constellation**, and the **Leo Star Constellation** and many others. Nabta Playa and the Giza Complex also served as Calendars and **Clocks**. The year 2011 movie entitled *"**Hugo**"* shows the intricate relationship between computers, **robotics**, and **clockwork**.

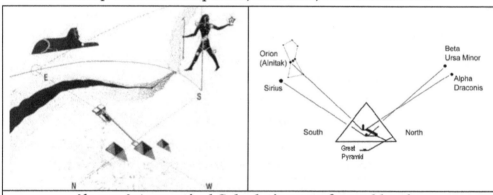

Above: Astronomical Calculations performed by the Giza Complex Analog Computer in Africa

In a book entitled *"**Mechanical arithmetic The history of the counting machine**"* by Dorr E. Felt, the writer describes how the world's first counting machine was developed based on a schematic developed by **North African Moors**. The book states that a French man named Gerbert d'Aurillac obtained a blueprint to develop a counting machine after having studied for years with the **Moors** at their Universities in **Cordova** and **Seville**.

In addition to the world's first Analog computer originating in Africa, the very word **"Computer"** also seems to have a relationship to Ancient African deities. The etymology and origin of the word "Computer" comes from the root word "**compute**", which comes from the Latin word "**computare**" meaning "**to count, sum up, reckon together.**" The Latin word "**computare**" is actually composed of the prefix "**com-**" and the suffix "**-putare**". The etymology and meaning of the words "**com-**" and "**-purare**" are:

- **com-:**   prefix usually meaning "with, together, **join**, or in **combination**" from the Proto-Indo-European word "*Kom-" meaning "beside, near, by, or with"

- **-putare:**   meaning "to pave, to count, to **think**, to consider, to **reflect upon**, to **reckon**, to trim, to prune, to strike, to **carve**, to **engrave**, to cut, to **separate**, **to open**"

The words "com-" and "-putare" which are the etymological origin of the word "Computer" are similar in phonetics and definition to the African Creation deities named **Khnum** and **Ptah**. The name **"Khnum"** is derived from the Ancient Egyptian root word "**khnem**" meaning "**to join**, to unite, **to combine**, and to build" which is also the same phonetics and meaning as the word "**com-**". The name **"Ptah"** has been translated as meaning **"Creator"** and also has been translated as meaning **"Opener"** or **"Sculptor"** with derivatives of the name **"Ptah"** found in the **Judeo-Christian Bible** as **"Japheth"** (Strong's H3315) translated as "opened", **"Riphath"** (Strong's H7384) translated as "spoken", **"Pathach"** (Strong's H6605) translated as "to open" or "to **carve** or **engrave**", "Pehter" (Strong's H6363) translated as "firstborn, firstling, or that which **separates** or **opens first**", and "Patsah" (Strong's H6475) translated as "**to part** or **to open the mouth**". Thus, the name **"Ptah"** is similar to the word "**-putare**" in phonetics and meaning. Therefore, the word **"Computer"** is similar in phonetics and meaning to **"Khnum-Ptah"**. Also, the deities **Khnum** and **Ptah** are found joined in Ancient Egyptian and African Cosmology in a variety of sources. Both Khnum and Ptah are called **"Father of the Gods"**. The book entitled **"Creation records discovered in Egypt: (studies in The Book of the dead)"** by George St. Clair notes that the **"7 Pygmy sons of Ptah"** are called the **"7 Khnemu"**. The "7 Khnemu Pygmy sons of Ptah" are said to be the source of the concept of the **"7 Arch Angels"** in religion and the **"7 Dwarfs"** in fiction. In the book entitled **"Egyptian Paganism for Beginners"** by Jocelyn Almond it states that the "Hymn to Khnum" attributes him with the title of **"Khnum-Ptah"** because both Khnum and Ptah were creator deities in Ancient Egypt.

KHNUM-PTAH

KHOM PUTAR

COM PUTER

COMPUTER

The etymology of the word **"Program"** comes from the prefix "pro-" meaning "before" and the suffix "-gram" meaning "word". Therefore, when we analyze the phrase **"Computer Program"**, we recognize that it actually breaks down to **"Com-Puter Pro-Gram"** which means **"With Ptah Before The Word"**. In the Memphite Theology of the deity Ptah, we know that prior to Ptah "speaking a word with his tounge" (**Hu**), he first had to "imagine it in his heart and mind" (**Sia**). The "word" or tongue of Ptah is likened to the **"logos"** (**logic**) and the "heart" or mind of Ptah is likened to **"nous"** (**reason**). Therefore, the "computer program", or that which was "with Ptah before the word" was the heart and mind (**Sia**) which provided the logic and reason to enable the automating, animating, and programming of creation. Thus, Africa is the origin of the world's first computer, Africa is the origin of the word "Computer" (Khnum-Ptah), and Africa is the origin of the philosophy of "Computer Programming" (with Ptah before the word).

0101 – ROBOTS AND CYBORGS IN AFRICA

Towards the end of the year **2009** movie entitled *"Transformers 2: Revenge of the Fallen"*, the **Robot** named "**Optimus Prime**" is depicted standing victorious next to the **Har-Em-Akhet "Great Sphinx of Giza"** in Africa. This scene is quite significant because it depicts the Past (the Great Sphinx) and the Future (Android Robots) of the same craft.

Above: Scene from the movie "Transformers 2: Revenge of the Fallen" with the Android Robot Optimus Prime standing next to the Har-Em-Akhet "Great Sphinx" in Africa

In order to comprehend how the "**Great Sphinx of Giza**" is related to **Robotics**, we must examine the philosophy and ideology of the culture which built the Great Sphinx of Giza and their reasons for building it. By comprehending some fundamental points, it is quite easy to see how Robotics started in Africa. In Ancient Egypt, there was a ceremony, ritual, and set of practices called the "**Opening of the Mouth**" ceremony. The oldest written evidence of the "Opening of the Mouth" ceremony is found inscribed on an artifact called the **Palermo stone** which dates back to about 2600 B.C. However, it is believed by many Egyptologist that the practices and rituals of the "**Opening of the Mouth**" ceremony that are inscribed on the Palermo stone could be much older than the artifact itself, and may date back to **Pre-dynastic African rituals**. The Egyptian **Medu Neter** term for the ceremony was /wpt-r/ which literally translated to "opening of the mouth". The root of the word /wpt-r/ also denoted an opening, splitting, dividing, separation, and determination of the truth (**binary**).

The "Opening of the Mouth" ceremony, as it is inscribed on the Palermo stone, describes a set of practices which were used to **animate statues**! The ritual of the "Opening of the Mouth" ceremony would take place in the Sculptor's Workshop or in the **Goldsmith's workshop** called the "**Hut-Nub**" (literally House of Gold). To the Ancient Egyptians, statues were made as vessels to house the "**energy**" of various deities. The "Opening of the Mouth" ceremony was the method which transferred the

Above: the Ancient African Egyptian Palermo Stone from 2600 B.C. The second line of the artifact describes how the "Opening of the Mouth" ceremony was used to <u>Animate Statues (create Robots)</u>.

energy of the deity into the statue, giving the statue a "**Ka**" (sometimes translated as "spirit") and giving the statue life, making the **statue animated**, and making the energy of the deity in the statue able to receive offerings. Recall that in the story of **Khnum**, it was the "**Ka**" that Khnum inserted into the newly created Human clay form which made the "clay Human" animated and "come to life". The "Ka" of Ancient Egypt is very similar to the concept called the "**All Spark**" in the "***Transformers 2: Revenge of the Fallen***" movie which caused inanimate metal objects to "come to life and transform". Therefore, when the **Har-Em-Akhet "Great Sphinx of Giza"** was built as a massive statue, it too would have received an "Opening of the Mouth" ceremony to give it a "Ka" and make it "animated" and "come to life". Thus, the **Har-Em-Akhet** and the philosophy of the "Opening of the Mouth" ceremony serve as a massive indicator that the concept of Robotics started in Africa. There is also an obvious relationship between the "Opening of the Mouth ritual" used to animate statues to the African Creation deity **Ptah**. Recall that Ptah created by "**opening his mouth**" to **speak** the "**word**" of creation. This "word" spoken by Ptah, is likened to the "**logos**" which is the etymological origin of the word "**Logic**" used in computers and robotics. Therefore, the "Opening of the Mouth" ceremony can be interpreted as a method used to place "Logic" into a statue to animate it which is synonymous with Robotics. Like designing and programming a Robot, the "Opening of the Mouth" ceremony was an extremely lengthy process with over 100 lines of code. Episodes 10 through 22 of the "Opening of the Mouth" ceremony dealt specifically with animating the statue. The "Opening of the Mouth" ceremony was performed when the statue could be positioned such that it would face the **Sirius Star Digitaria** during the rising of the **Orion star constellation**. To the **Dogon** people of **Mali West Africa**, the Sirius Star **Digitaria** or **Sirius B** is called "**po tolo**" and is considered the origin of all life. There were usually several people involved in Animating the Statue during the "Opening of the Mouth" ceremony. **Tehuti** was said to be the symbolic supervisor of the project. **Sokar** was said to symbolically open the eyes of the statue and **Ptah** was said to symbolically open the mouth of the statue. The mouth was opened by striking the statue with a tool called the "**Adze** of **Anupu**" which was shaped like the "**big dipper**" **Ursa Major star constellation**.

It was not until around 2300 B.C. that the "**Sem**" funerary priests started using the "Opening of the Mouth" ceremony to re-animate the mummified bodies of deceased Humans. Therefore, the concepts of "**Resurrection by way of Technology**" and "**Transhumanism**" originated with the Ancient Africans in Egypt also. Inscriptions on the walls of the temple of Edfu describe how the "Opening of the Mouth" ritual allowed the mummy to eat, breathe, see, hear, and experience life again. When the "Opening of the Mouth" ceremony

A NUPU
Ancient Egyptian deity of the Afterlife and performer of the Opening of the Mouth Ceremony

started being used to re-animate the dead, the **ritualistic sacrifice of animals** began to be incorporated into the ceremony. Not only was the "Opening of the Mouth" ceremony used to give life and animate Statues and deceased bodies, but it was also used to **give life to buildings, temples, and rooms**. Therefore, the idea of a "**living building**" or "**Virtual Reality**" is another concept which originated with the Ancient Africans in Egypt.

It is reasonable to ask "did the 'Opening of the Mouth' ceremony actually animate the statues and mummies, or did the Ancient Egyptians merely imagine that it did". Considering that the ritual was performed on Statues before it was ever used on a person, if the ritual did not work, the Ancient Egyptians certainly must have believed that it worked in order to start using it on deceased people, family members, loved ones, and Rulers. It is quite possible that prior to using the "Opening of the Mouth" ceremony to animate deceased bodies, the body of the deceased was **cremated** and the **ashes** were used to create a **statue** that was then animated by way of the "Opening of the Mouth" ceremony. It is reasonable to ask the question "if the Great Sphinx did indeed receive an 'Opening of the

Mouth' ceremony and the 'Opening of the Mouth' ritual actually works, then why isn't the Great Sphinx walking around animated like a Robot now?" One possible explanation is that to the Ancient Egyptians, when a **statue's senses (robot's sensors)** are destroyed, then the energy of the "Ka" leaves the statue. Thus, since the Great Sphinx has had its **nose destroyed** and **eyes destroyed**, the animating "Ka" energy that was imbued upon it during its "Opening of the Mouth" ceremony has left the statue and subsequently the statue can no longer move around animated like a Robot. During the later periods of the Egyptian empire, the use of glass for eyes of statues became prevalent to make the statues seem more lifelike perhaps because the knowledge and ability to actually animate the statues and bring the statues to "life" had been lost.

The process of creating "Robots" in Africa, that is to say, creating statues and ceremonially and ritualistically inserting energy within the statue in order to serve a purpose, has been prevalent throughout African culture over time and is most commonly likened to some form of "**magic idol**" or "**fetish**". In the **Congo** region of **Central Africa**, statues called **N'Kisi** are created to house "spirits" or energies which perform certain tasks including healing, communication with the dead, hunting, business, and sex. The N'Kisi "Robot" statues are sometimes considered idols and charms. A particularly aggressive version of the N'Kisi is called the

Nkisi Nkondi, African Congo "Nail Fetish" Robot

N'Kondi which means "**to hunt**" (a "**seek and destroy**" type of robot). The N'Kondi "Robot" statues are created to hunt down and attack liars, thieves, killers, and wrong does. The N'Kondi statues are "**activated**" by hammering nails into them and thus are sometimes referred to as "**Nail fetishes**". The N'Kondi "nail fetish" statues are the origin of the concept of "**Voodoo Dolls**" that are "activated" by placing pins into the doll's body. The traditional African philosophy, culture, science, spiritual system, and practices that incorporated the creation of Robotics in the form of statues, dolls, and puppets has spread all over the world and appears under many different names. **Hoodoo** is one manifestation of the tradition called "**folk magic**" by people of African descent in the Americas. **Voodoo** is another manifestation of the tradition by people of African descent in New Orleans, Haiti, the Caribbean Islands, and South and Central America. **Vodun** is another manifestation of the tradition by Africans in West, Central, and South Africa.

One of the biggest differences between Hoodoo and Voodoo, is that Voodoo is a religion, whereas Hoodoo is just a set of practices or a collection of "techniques" (technology). Other manifestations of the tradition have been called by the names **Palo**, **Santería**, and "**Rootwork**". The creation of what is commonly called "**Voodoo dolls**" is found in New Orleans Voodoo, and also in Hoodoo practices of creating "**poppets**". The word "poppet" is derived from the word "**puppet**", and "**Hoodoo Poppets**" are dolls made to represent a person and used to **program** or "**cast spells**" on the person whom it represents. Much like the "**Voodoo Doll**", the "**Hoodoo Poppet**" is a doll which serves as a "**remote control device**" which can be used to control the movements of a living person. Therefore, there is a clear and obvious relationship between "**Voodoo Dolls**", "**Hoodoo Poppets**", and "**Remote Control Robots**". The etymology of the word "**Puppet**" comes from a word meaning "**person whose actions are controlled by another**" and shares the same etymological origin as the word "**Pupil**", which shows how being a "Pupil" or **student in school** is like being **programmed** or controlled like a puppet or robot.

The use of Robot puppets in African culture occurs for a variety of purposes. African puppets are used to tell stories, entertain, and pay homage to deceased ancestors. Some of the African tribes who utilize "Robot" puppets in their culture include the **Bamana** people of **Mali**, the **Ibibio** and **Ekoi** people of **Nigeria**, and the **Nyamwezi** people of **Tanzania**. Also, the prevalent use of **African Masks** throughout the various African cultures is a manifestation of the same idea that an inanimate object can be "animated or brought to life", or that a certain "energy", "spirit", or "god" can embody a person or object. The **"Mask of God"** is made to speak, dance, and move about acting as an **"Avatar"** during traditional festivals and ceremonies. Some of the notable Mask artifacts include the **Bronze Masks** of **Benin**, the **Gelede headdress** of the **Yoruba** people in **Nigeria**, the **Sibondel Headdress** of the **Baga** people in **Guinea**, the **Baluba Mask** from the **Congo** region in Central Africa, and the **Kanaga Mask** worn by the **Dogon** people of **Mali**.

West African Mask

Baluba Central African Mask

The relationship between "puppets" and "robotics" was show in the year 2001 science fiction movie entitled "***A.I. Artificial Intelligence***" about a "robot made to look like a child" who wanted to be a **"real boy"** which was reminiscent of the year 1940 Disney animation film entitled *"Pinocchio"* about a **puppet who "came to life"** and desired to be a "real boy". The movie *"Pinocchio"* also has several occult symbolisms in the name of the characters. The name Pinocchio comes from the Italian words *"Pinolo-"* meaning **"pine nut"** and *"-occhio"* meaning **"eye"**. Also, the name of the creator of the Pinocchio puppet in the movie was **"Gappetto"**, which comes from the Hebrew word **"Japheth"** or "YAA-PATH" (Strong's H3315) which as we mentioned earlier is derived from the name of the African Creation deity **"Ptah."**

Throughout African culture, the creation of "robots" or objects to house certain "energies", and the objects themselves, such as Masks, Puppets, Statues, Dolls, charms, amulets, and other objects have been called a variety of names. These **African robots** have been called "**Juju**" coming from a **French** word meaning "**toy**" and has been used to refer to various objects or fetishes used

"Fetish" Robot statues from the animated movie entitled "Kirikou and the Sorceress"

by African tribes which have the ability to house certain **"energies"** and **"powers"**. **Juju** is also called **Bò** or **O bò** in Yoruba. In the Twi language of the Akan people in West Africa, these African Robots are called **"Abosom"** which refers to energies which can exist within objects. These African robots have been called **"Ouangas"** or **"Talisman"** coming from a Greek word meaning "to initiate into the mysteries" and is used to refer to **inanimate objects that one makes animate**. In the year 1998 animated film entitled *"Kirikou and the Sorceress"*, the sorceress named **Karaba** commands an army of **wooden robots** that are referred to as "fetishes". These African robots have also been called **"Obia"** which is a term used to refer to any object, talisman, or lucky charm that is created and used in folk magic and "sorcery". Obia was used by descendants of African slaves in the Americas against their captors. The word Obia can be traced to a variety of sources including the Igbo word *"dibia"* meaning *"healer"* and the Ashanti word *"Obayifo"* meaning **"witchcraft"**. In the year 1988 horror movie entitled "Child's Play" a **toy doll comes to life** when a **practitioner of Voodoo** transfers his **consciousness** into the body of the doll. The ability to transfer one's consciousness into a doll or machine using African Voodoo practices is similar to the concept of **"Mind Uploading"** into computers and robots currently being discussed in conversations about Transhumanism in the fields of Science and Technology. It is also interesting to note that the word **"O.B.I.A."** is used as an abbreviation in computer science for **"Object Based Image Analysis"** which is the ability to extract information from images using digital image processing software. "Obia" or "Object Based Image Analysis" is the computer software required for electronic facial recognition software and also to read bar codes and QR codes.

Much like the "Opening of the Mouth" ceremony in Ancient Egypt was used to animate both statues and deceased bodies of mummies, the rituals used to animate statues and dolls in the African traditions of Hoodoo, Voodoo, Vodun, Palo, Santería, and "Rootwork" are also used to animate dead bodies or "create

Zombies". In Trinidad and Tobago, one form of **Obia** is a "**stilt dancer**" called "**Jumbie**" in which the Jumbie stilt dancer mimics the movements of a **Zombie**. Although the word "Zombie" is usually used to refer to an animated corpse, it can also be used to refer to a living person in a hypnotic dreamlike trance whose is being manipulated using some form of **mind control**. Living people are turned into "Zombies" through the intake of some form of Drug. In year 2012, there was an epidemic termed the "**Zombie Apocalypse**" due to the number of people using drugs laced the "**bath salts**" which caused them to act with Zombie-like behaviors. The story of "***Frankenstein, the Modern Prometheus***" is an example of the relationship between creating **Zombies** and **Robotics**. In Voodoo, **Zombie Robots** are created to perform a specific task. In **South Africa**, the Zombie "spell can be broken" or the "**programming can be overridden**" by a traditional healer called a **Sangoma**.

The word "**Zombie**" is actually African in origin. In Haitian Vodou, "**Zombi**" is another name for the Loa energy "**Damballah**". In Haitian Vodou, Damballah is a **sky deity** and is considered the **primordial serpent of life**, animation, and ruler of the mind and intellect. In the year 1988 horror movie entitled "**Child's Play**" a Voodoo practitioner calls on Damballah to transfer his consciousness and soul into the body of a toy doll. The **Vodou veve symbol** for Damballah with two serpents is similar in appearance to the "**compass and square**" of **Freemasonry**. Damballah is known by the name **N'Haala** the **sky deity** to the **Balanta** people of **Guinea-Bissau** in West Africa. Damballah is known as **Obatallah** in **Nigeria**. In the **Yoruba** tradition of Nigeria, Obatallah is the creator and shaper of human bodies and is much like the deities **Khnum** from Ancient Egypt and **Prometheus** from Latin mythology in this regard. Obatallah is considered the owner of all the "Ori" or "heads" because it is believed that the

"soul resides in the head". The origin of the word "Zombie" actually comes from a central African Congo word **"Nzambi"** which has been translated to mean "god" but actually transcends the definition of the word "god". **Nzambi** is akin to names found in a multitude of traditional African systems including **Njambe, Nyambe, Nyame, Nyamma, Amba, Amma, Ama, Amit**, and **Amun**. Nzambi refers to the life, vitality, and the energetic source of all creation, movement, and Thermodynamic changes and exchanges of **Energy** in Nature. All things, living or dead, animate or inanimate, have the potential to undergo change, movement, and animation in Nature. Nzambi is a form of **African Creation Energy**, and the utilization and manipulation of this energy by African people is part of our traditional **Animist** way of life.

Overtime, the sciences of **Voodoo, Hoodoo, Vodou**, and **Vodun** have been used for negative purposes and have decreased in reasonability which has led to a negative connotation being associated with these sciences. It is important for African people to restore the positivity and reasonability back to Voodoo, Hoodoo, Vodou, and Vodun so that our traditional African sciences can aid us in our Liberation and freedom. If working with Negative energies or negative "spirits" was indeed beneficial, then these energies would not have allowed the enslavement and colonization of African people. Instead of working with Negative energies, African people should work with positive energies which will empower African people to restore us to our greatness as Royalty, Elites, and Sovereigns. Using sound logic and right reason after experiencing evidence to determine the effectiveness, applicability, and practicality of any claim or any science is paramount. If evidence has been experienced for African people to have reason to know that the claims of the sciences of Voodoo, Hoodoo, Vodou, and Vodun do in fact work, then these sciences should be taught to all

African people who can comprehend the concepts, in the same way that computer science is taught to any student who can comprehend the concepts, to show that the effectiveness of the science is indeed replicable and yields results. However, if these sciences have degraded to the point where it is truly only deception, ritual, and a big show, then African people must restore positivity and reasonability to make our traditional sciences substantial and verifiable. A science is verifiable and replicable when it yields the same results every time the conditions are the same.

Many of the traditional African ways of life have been labeled under the umbrella term "**Animism**" which is the acknowledgement that the world of "**energy**" is intertwined with the material world, and therefore the energy that makes "life" possible is in everything: people, animals, plants, rocks, dirt, water, minerals, wind, air, fire, etc. The word "Animism" comes from the Proto-Indo European word "*ane-*" meaning "**to blow or to breathe**" and is also the etymological origin of the word Animal, **Animation**, and one of the earliest forms of robotics called **Animatronics**. The science of Animatronics was used to create one of the earliest modern robots called an **Automaton**. It is interesting to note that the word **Automaton** is phonetically similar to the names of two of the principles related to the Ancient Egyptian Sun deity **RE** as "**Atum-Aton**". Animists not only pay homage to the **energy** and **creative force** of life, but also utilize that energy to make things become animated; this is why the creation of Robots is an operative expression of traditional African Animist culture. The African religion of Vodun is classified as an Animist religion because the word "Vodun" comes from a word in the Gbe family of languages spoken in West Africa which is translated as "**spirit**" but more closely refers to the "**energy**, **force**, and **power** which governs the Universe and everything within it." From a traditional African Animist perspective, Robotics and science are comparable to the study of Nature and the energy of the Universe. However, from a perspective and mentality programmed by the three major Monotheistic "Abrahamic" religions (Judaism, Christianity, and Islam) Animism, Robotics, and Science are all "evil" and "pagan" because the ideas within these fields challenge the religious definition of Spirit, Soul, "free will", and also challenges the very definition of what it means to be "human". Firstly, in the "Abrahamic" religions, the concept of **"Free Will"** is supposed to be a characteristic unique to Human beings. Therefore, the idea of an **Autonomous robot** or **Automaton**, with the word **"Auto-"** meaning **"acting on its own will"**, is a direct contradiction to the religious doctrine about "free will being unique to humans". Furthermore, the concepts of "Spirit" and "Soul" as presented in the "Abrahamic" religions, or as comprehended colloquially, does not stand up to scrutiny. The table below presents a comparative analysis of the definitions of "Spirit" and "Soul" from the original language of the "Abrahamic" religions.

Abrahamic Definition of Spirit and Soul	
SPIRIT	SOUL
Strong's H7307 - ruwach	Strong's H5315 - nephesh
air, gas, wind, breath	that which breathes, the breathing substance
disposition	self, creature, person,
animation, vivacity, vigor	activity of mind, activity of the character, seat of the appetites
emotional impulse: temper, anger, courage, sorrow, etc	desire, emotion, passion, activity of the will
energy of life	life, living being
mind	mind

As we can see from the above definitions, the meaning of the words "Spirit" and "Soul" are basically indistinguishable in the original language of the "Abrahamic" religions. In addition to the actual meaning of the words "Spirit" and "Soul", there are also other definitions that have been colloquially attributed to "Spirit" and "Soul" which most people associate to the words "Spirit" and "Soul" in their mind. There is an observable disconnect between the actual meaning of the words "Spirit" and "Soul" in the original language of the "Abrahamic" religions, to what people think and associate with the words "Spirit" and "Soul" in their mind.

| Colloquial Definition of Spirit and Soul ||
SPIRIT	SOUL
Immaterial and Intangible	Immaterial and Intangible
survives after the body dies	Immortal
	Belong only to Humans
consciousness, intelligent or sentient part of a person	mind, psyche
personality; the feeling, quality, or disposition characterizing something	self, person, individual
seat of emotions, mood	moral and emotional part of person
animating or vital principle in man and animals, vigor, to make more active or energetic	animate existence, actuating cause of an individual life
an alcoholic liquid solution of a flammable substance	etymology: coming from or belonging to the sea
etymology: to blow or breathe, breath, pneuma	
able to travel and posses people, places, and things, and can appear as a ghost	

The multiple definitions of "Spirit" and "Soul" logically contradict each other individually. If the original word for "Spirit" (**ruwach**) and the original word for "Soul" (**nephesh**) both deal with "air, wind, and breath", then why is it that only the word "*Spirit*" itself actually comes from a word meaning "**air, wind,** and **breath**" whereas the word "*Soul*" itself comes from a word meaning "**coming from the sea**"? If the original word for "Sprit" and "Soul" both deal with "air, wind, and breath", then why is there the belief that either "Spirit" or "Soul" can only belong to Humans; don't animals breathe too? If the original word for "Sprit" and "Soul" both deal with "air, wind, and breath", then why is it believed that a person's individual "air, wind, and breath" will survive after the person dies and stops breathing? If the original word for "Sprit" and "Soul" both deal with "air, wind, and breath", then this means that "Sprit" and "Soul" cannot exist in a vacuum environment void of gases and therefore cannot travel through outer space. If "Spirit" and "Soul" are both "air, wind, and breath" then "Spirit" and "Soul" cannot be "Immortal" because if all of the gas is removed from wherever the "air, wind, and breath" of the "Spirit" and "Soul" is, then the "Spirit" and "Soul" will cease to exist. If the original word for "Sprit" and "Soul" both deal with "air, wind, and breath", then how is it that "Sprit" and "Soul" is considered "Immaterial and Intangible" because "air, wind, and breath" are all material and tangible? If "Sprit" and "Soul" is "Immaterial and intangible" then how was it ever detected for it to be known that such a thing as a "Sprit" and "Soul" exists? If a "Sprit" can appear to someone, then why does the "Sprit" always appear wearing clothes; are clothes part of Spirit, and if so, how and why? Since "Sprit" and "Soul" share so many common definitions, why is there the need to distinguish two separate terms? The questions, contradictions, and logical fallacies stemming from the concepts of "Spirit" and "Soul" are endless.

The result of attempting to synthesize and **scientifically justify the doctrine of "*Spirit*" and "*Soul*"** using the term "**Energy**" yields the following sequence:

1. **Energy** is fundamental throughout the universe and cannot be destroyed but merely transformed, i.e. **Immortal**

2. **Energy** exists within gases like the **air** and **atmosphere**

3. **Air** can be **breathed** in to obtain **Energy**, giving a living thing **life**, **vigor** and the ability to move and be **animated**

4. The **energy** obtained from the **air** by **breathing** will provide energy to the brain, **mind**, and mentality

5. Utilizing **Energy**, the **mind** determines a person's **personality**, **character**, **emotions**, and **moral decisions**

6. Utilizing **Energy**, certain **mental abilities** like "**abstract thinking**" and **higher order reasoning** seem to be attributes found only in **human beings**

7. The **Energy** that once powered a person's body and mind will still exist, and "**survive after the person dies**" when the energy is transferred or transmitted to another substance

8. Therefore, the **Energy** that once powered a person's body and mind could indeed be transferred to another person, place, or thing

9. Utilizing **Energy**, the works, deeds, and actions a person performs while they are living, can **live on after the person dies** for the person to be **Remembered Always**

Since "Energy" is verifiable, religionist want to attribute "Spirit" and "Soul" with Energy, however religionists are unwilling to deal with what the scientific realities of Energy will mean for their doctrine. For starters, the religious doctrine of an "**afterlife**" in "**Heaven**" or "**Hell**" in a **garden** or **eternal fire** makes absolutely no sense in the scientific study of Energy.

Without the doctrine of being punished or rewarded after the person is dead, religion loses its ability to get people to follow the instructions of its program. And so although Religionists want to equate "Spirit" and "Soul" with Energy, if they are to do so, the unverifiable and illogical concepts associated with "Spirit" and "Soul" must be abandoned. If you accept the realities of Energy, then you comprehend how the flow of Energy can make a thing become activated, animated, and "move" like it is alive as in the field of Robotics. If you accept the realities of Energy, then you comprehend how Energy can flow within certain substances like "semiconductors" or "brain matter" and store energy in the form of information to create what we call **memory**. If you accept the realities of Energy, then you comprehend how Energy can flow within certain substances like "semiconductors" or "brain matter" to produce logic, reason, and "mental process" which then can be programmed to create emotions, personality, and make decisions. If you accept the realities of Energy, then you accept the realities of Science, Computers, and Robotics and you accept the reality that all of the attributes which Religionists associate with "Spirit" and "Soul" can be described using Energy, and Energy can flow and create "life" animation, logic, and reason in a multitude of substances without necessarily having to use "**air**" and "**breathing**". One of the main reasons why people fear computer science and robotics is because it forces Humans to realize that we are not as "**special**" as we were originally programmed to think. And so the old illogical and unreasonable program built on Religious concepts about "Spirit" and "Soul" is outdated, and we must adopt a new Program based on "**Sound Right Reason**" to run on the operating system of our mind. The mind determines the tasks and jobs a person performs and also determines the person's personality, emotions, and decisions. These attributes associated with the mind are usually likened to the antiquated concepts of "Sprit" and "Soul". These attributes associated with the mind are also similar to a Computer Program since a "Program" is **"intangible"** and can be written to determine specific tasks, emotions, and decisions. **"Spirituality"** is defined as the "belief" in an **immaterial reality** and **a set of codes which the person lives by**. Thus, we can liken "Spirit" and "Soul" to the **"Program and Controller"** of people. Computer Programs are made out of Energy, and Energy is an **"Immaterial Reality"** that you do not have to "believe" exists, but rather you can verify it exists with evidence, experience, and reason. Since "Computer Programs" are created by mental concentration, we can also assume that "Sprits" (People Programs) were also created through mental concentration. Therefore, you must ask yourself, "Who created the Program (Spirit) that is controlling me, and was this program (Spirit) created to serve my best interest or someone else's best interest?"

Opponents of Computer Science and Robotics who are consciously or unconsciously proponents of Religious programming will tell you to embrace "Spirit" and "Soul" which are both unverified and will try to **encourage you to embrace Death** and **HOPE** for a "Spiritual" **eternal afterlife** in "Heaven". Meanwhile, other groups of people who have embraced derivatives of your traditional African culture are creating "Heaven" and "Eternal Life" for themselves **RIGHT NOW while they are living on Earth** using Computer Science and Robotics. Opponents of Computer Science and Robotics who are consciously or unconsciously proponents of Religious programming will tell you to **"Hope"** and **"Have Faith"** that some "Spirit" will **come and save you**, and this philosophy only programs your mind to become a lamb to the slaughter while people who are **not waiting** and **not hoping** are **doing** what needs to be done to ensure the best possible outcome for their interest.

So as we accept the reality of Energy, we recognize that energy can flow through or be transferred to anything, therefore in the religious sense, **"anything can have a spirit (energy)"**. If the energy can be made to perform the logic processes of reason as is the case with semiconductor devices, then in the religious sense, **"anything can be given a Soul (energy to perform logic and reason)"** as well.

We find that the Animist Ancient African Egyptian concepts of **Ka**, **Ba**, and **Akh**, which are translated as "aspects of the spirit and soul", share many commonalities with the flow and use of energy in the fields of Robotics and Computer Science. In Ancient Egypt, the **Ka** is considered the **"spark of life"** and vital essence which energies, activates, animates, and gives something "life". The Ka was symbolized as two raised arms. The concept of **"HeKa"** which is translated as "Magic" but is more closely related with the concept of "Engineering" was the activation and use of a person's Ka or life force energy in words, deeds, and actions to create various technologies. When the Ka left a person's body, the person was dead. In Ancient Egypt, the **Ba** is considered a person's personality, character, and moral decisions which are all determined by mental processes. To some regard, the Ba is like the person's programming. The Ba was symbolized by a bird with a human's head. The "Ba" was believed to live on after the death of the person in the way the person was remembered by others who met and dealt with the person while they were living. After the death of the person, the Ka and the Ba were said to unite. The unification of the Ka and the Ba created the **Akh** which was a manifestation of the person's **effectiveness** which would live on after death. The interpretation of the Ka uniting with the Ba to create the Akh is basically a way of saying "*it is the works you do while you are living (Ka) combined with your personality, the way you treat people, and the way* *you are remembered (Ba) which will determine how effective you were in life (Akh) and if you will be remembered always (live forever).*" When the Ka and the Ba united to form the Akh, it was believed the Akh would ascend to the sky to become a star. The Akh was also symbolized by an Ibis bird.

3 ASPECTS OF THE HUMAN USE OF ENERGY (Commonly called "Spirit" and "Soul") IN ANIMIST AFRICAN EGYPTIAN THEOLOGY		

KA Symbolized by Raised Arms	**BA** Symbolized by a Bird with the head of a Human	**AKH** Symbolized by an Ibis bird which becomes a Star

The Ancient Egyptian animating concept called Ka, mental concept called Ba, and effectiveness concept called Akh that were considered "aspects of Spirit and Soul" in Humans are very similar to the flow of energy in Robots. Electrical energy flows through the robot to make it move and be animated (Ka). Electrical energy flows through the robots microprocessor so that the robot can use logic, reason, and make decisions and judgments based on its programming (Ba). When the energy used to animate the Robot (Ka) and the energy used to program the Robot (Ba) actually enable to Robot to do what it was created to do, then the Robot is considered effective (Akh). Therefore, comprehending the flow of energy through a Robot and its relationship to the Ancient Egyptian Ka, Ba, and Akh concepts that are considered aspects of "Spirit and Soul", then it is fair to conclude that Robots and Computers can indeed have a "Spirit" and "Soul".

In the Ancient Egyptian **"Book of Coming Forth by Day"** (commonly called the **"Book of the Dead"**) the **"Ba"** (mental aspect) of the Sun deity **RA** was called the **Bennu Bird**. The Bennu bird can be likened to the **Phoenix** or **"fire bird"** found in cultures around the world symbolic of **resurrection**, **rebirth**, and **regeneration**. In the cosmology from the city of Heliopolis (Annu) in Ancient Egypt, the Bennu bird was said to have rested upon the **Primordial mound** (**Ptah**, also pronounced **TA**) that rose from the **primordial waters** of **Nu**. It is interesting to note that the principles related to the Bennu bird (**RA-BA**), symbolic of resurrection and rebirth, which sits on the primordial mound called **TA**, can be combined as **RA-BA-TA** which is phonetically similar to the word **"ROBOT"**, a modern symbol of resurrection and rebirth. Also, in the **"Hymn to Khnum"** from the city of Elephantine, Khnum is equated to the **"RA-BA"** (the BA of RA) and also Khnum is equated to the deity **Ptah** as **Ptah-Tatenen**, the **"Creator of Creators"**. When we combine these three attributes of Khnum, we get **RA-BA-PTAH** also pronounced **"RA-BA-TA"** which is phonetically similar to the word **"ROBOT"** and in the cosmology related to Khnum, he was the creator, designer, and activator of the **Human "Robot"**.

The flow of energy to be used for life and mental processes appears in a multitude African philosophies. In books written by the philosopher and scribe named "Afroo Oonoo", the "spirit" or energy which animates and gives "life" is called "ZoopooH", and the "spirit" or energy which performs the mental reasoning processes is called "NoopooH". In West African culture, "Spirits" or "Energies" are the "Programs" which determine a person's personality traits. For example, in the **Akan** culture of West Africa, the **"Kra-din"** or "Soul Names" are names given to people based on the day of the week on which they are born. Each day of the week corresponds to a different Celestial Body which is believed to be an Abosam in the Akan culture and sends energies to program the person's personality and characteristics. The Kra-din can be likened to a person's default programming which occurs at birth.

Day of Week	Celestial Body	Abosam	Name
Sunday	Sun	Awusi	Kwesi (Male), Akosa (Female)
Monday	Moon	Dwo	Kwadwo (Male), Adwo (Female)
Tuesday	Mars	Bena	Kwabena (Male), Abena (Female)
Wednesday	Mercury	Aku	Kwaku (Male), Akua (Female)
Thursday	Jupiter	Yaw	Yaw (Male), Yaa (Female)
Friday	Venus	Afi	Kwafi (Male), Afia (Female)
Saturday	Saturn	Ama	Kwame (Male), Amma (Female)

In West African culture, additional "Spirits" or Energies are also able to program people after birth. When a "Spirit" or "Energy" programs a person, it is considered a **"possession"**, **"incarnation"**, or **"channeling"** and the result of this can be either positive or negative. In West African culture, certain people are trained to master and control one or more "Programs" (Spirits) running on their operating system (Mind and Body) making them **"Priests"** and **"Priestesses"**. One example of a West African Spirit, Energy, or Program which is positive and productive is the Abosam named Nana Adade Kofi which is a program that enables people to be a blacksmith and work with iron, metal, and create technology. The **Nana Adade Kofi** Abosam program is similar to the Yoruba Orisha **Ogun** program which is the original source of the **Iron Man** concept.

The West African Akan **Kra-din** "Soul Names" are similar to the concept of **Zodiac signs** in **Astrology** which suggests that **celestial bodies** like **planets** and **stars** can control, influence, and **program** the behavior and characteristics of people. The belief that Celestial bodies program Human "Robots" is common in a variety of Animist traditions found around the world. The **Shēngxiào** is a form of Chinese Astrology which uses a 12-year cycle to relate each year to an animal and people born within that year are believed to have attributes of the animal. The creation of Robots is also found in Ancient Asian culture. The Daoist text called the *"LieZi"* tells a story about the creation of a Robot in Asian culture occurring around 900 BC. In the story, an engineer named **Yan Shi** presents a life-sized, human-shaped Robot made of leather, wood, and glue to **King Mu of Zhou**. The Robot was painted red, black, white, and blue and could walk, move its head, sing, and even "flirt" with the women in the audience. When King Mu of Zhou saw the Robot flirting with the women, he became angry, and in order to get the king to settle down, the engineer named Yan Shi had to take the Robot apart to show the king that the Robot is indeed artificial. The ancient history of Animism, Automatons, and Robots in Asian culture has motivated modern day Asians to take part in the operative expression of their culture in the field of Robotics by creating some of the World's top Humanoid Robots including the **Asimo** robot from the **Japanese** company **Honda**, the **HRP** (Humanoid Robotics Project) Robot from **Japan**, the **KHR** (Kondo Humanoid Robot) from **Japan**, and the **Biolid** and **DARwIn-OP** robots both from the **Korean** company **ROBOTIS**.

| **Asimo** | **DARwIn-OP** | **KHR** |

Since African Animist traditions influenced the creation of the "Abrahamic" religions of Judaism, Christianity, and Islam, there is evidence of "Animism" and Robotics that can be found in the "Abrahamic" religions as well.

In Hebrew mythology, a **"Golem"** is a type of Robot created by **Jewish Rabbis**. The Golem is inanimate matter, usually made of mud or clay, which is shaped into a Human form and brought to "life" and animated to perform labor and various tasks for the Rabbi. The Golem could be activated by inscribing the Hebrew word *"emet"* (אמת, meaning **"truth"** in Hebrew) on the Golem's forehead, and deactivated by removing the letter *aleph* (א) from the word *"emet"* which would change the word inscribed on the Golem's forehead from "truth" to **"death"** (*met* מת, **"dead"** in Hebrew). The

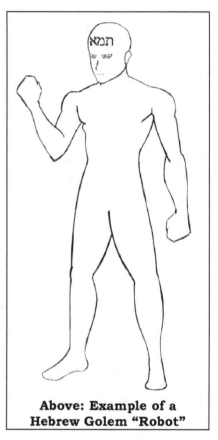

Above: Example of a Hebrew Golem "Robot"

Golem Robot could be **programmed** by placing paper with written instructions into the Golem's mouth. The word "Golem" appears in the Bible in **Psalm 139:16** in the statement *"Thine eyes did see my substance, yet being unperfect (golem - Strong's H1564)."* The word "Golem" as used in the Bible for an **"unperfected substance"** is also used to refer to a Human **embryo** or **fetus**. In the Jewish Talmud, the first Human named **Adam** was originally created as a Golem when he was "**created from Clay**". The story of the first Human Adam being created from clay in the Abrahamic Religions obviously shares similarities to the Latin story of **Prometheus** creating "Man" from clay and water, and also shares similarities to the Ancient Egyptian story of **Khnum** creating the first Humans from clay. The etymology of the word "**Robot**" comes from an old Slavic word "*Rabu*" which meant "**slave**" and the word "*Rabu*" is phonetically similar to the Hebrew words "*Rabbi*" meaning "**my master**" and "*Rabbo*" meaning "**his master**". Studying the etymology of the words, it is interesting to note that the Jewish **Rabbis** (masters) built **Robots** (slaves) called **Golem**. In the Abrahamic Religions, since God created Human beings "**in his image and after his likeness**" then the ability to create "life" like God is one of the abilities that should have been bestowed upon Human beings. Therefore, the creation of **Golem Robots** by **Jewish Rabbis** was considered a sacred practice and a way of worshiping God by becoming closer to God and making use of the power of God bestowed upon Humans to create "life" in the same way God created.

Stories, cosmologies, and mythologies about "statues coming to life" and ancient Automata and ancient Robots can also be found in Ancient European culture and have served as motivation for modern day European and North American Robot designers. The ancient Latin poem entitled "*Metamorphoses*" discusses a sculptor named **Pygmalion** who fell in love with an ivory statue of a woman he created named **Galatea**. In the story, Pygmalion prays for the statue to come to life, and when the statue comes to life, Pygmalion and Galatea conceive a child named **Paphos**. In Greek mythology, there are several stories of Robots and "statues coming to life" including the story of **Talos**, the bronze robot who protected the city of Crete from invaders, and the story of the craftsman

THE AFRICAN INITIALIZATION OF COMPUTER SCIENCE

named **Daedalus** who is able to create talking statues, and the story of the Greek blacksmith deity of technology named **Hephaestus** who builds a robot to assist him in his workshop and also created a female Robot out of clay named **Pandora**. The book entitled "**P.T.A.H. Technology**" by **African Creation Energy** describes how the Greek deity **Hephaestus** was derived from the Ancient African Egyptian deity **Ptah**. The book entitled "*Speaking Machines*" describes how in the 13th century, **Albertus Magnus** and **Roger Bacon** built **talking androids** which were destroyed by **Thomas Aquinas** out of fear. The Ancient History of Robots and Androids in European culture has motivated modern day Europeans and North Americans to continue with the tradition by creating robots including the **NAO** robot developed by **Alderbaran Robotics** of **France**, the **REEM** robot developed by **PAL Robotics** of **Spain**, and the **Zeno Boy Robot** developed by **Hanson Robokind** of **America**. The company Hanson Robokind uses the symbol of the Egyptian "**all seeing eye**" in their company logo, and also the **LAMI laboratory** of **Switzerland** has developed a robot named "**Khepera**" which comes from the Ancient African Egyptian deity depicted as a **scarab beetle** name *Kheper* meaning "**transformation**". It is obvious that the Ancient African traditions have travelled all over the world and have been used by other groups and races of people as motivation in the field of computer science and robotics. Africans and people of African descent must reclaim our philosophy and heritage and take part in the operative expression of our culture in the field of Robotics and Computer Programming.

There are several projects taking place worldwide that aim to be the catalyst in African computer science and robotics development. The **African Robotics Network (AFRON)** was established in April 2012 by **Ayorkor Korsah** and **Ken Goldberg** as an initiative to enhance robotics education, research, and industry n Africa. AFRON's advisory board includes **Tim O'Reilly** and **Dale Dougherty** co-founders of **O'Reilly Media** and **Make Magazine**. AFRON was inspired by the **European Robotics Network (EURON)**, but while EURON is focused on Robotics Research, AFRON is focused on education and industry. AFRON co-founder **Ayorkor Korsah** is a professor at Ashesi University in Berekuso Ghana West Africa. **Ashesi University** was founded by **Patrick Awuah** in 2001 and is considered Ghana's up and coming "**Silicon Valley**". **CODE2040** was founded by **Tristan Walker** in 2012 to give Black and Latino University students internship opportunities with Silicon Valley companies. "**Black Girls Code**" was founded in 2011 by **Kimberly Bryant** to motivate Black females to participate in the creation and development of computer programs, engineering, and technology.

In year 2010, a student named **Sam Todo** from **Togo, West Africa** developed an android Robot called **SAM10** out of old TV parts. The SAM10 robot is able to walk, greet people, and calculate the distance of objects in front of it in order to avoid obstacles.

Sam Todo of Togo West Africa, creator of the SAM10 Robot

Sam Todo developed the SAM10 robot to provide motivation and inspiration for other Africans to study computer science, robotics, mathematics, and engineering.

0110 - CYBORGS IN AFRICA

A **Robot** by definition is an electro-mechanical device that is designed to perform a task either automatically and autonomously, or semi-autonomously as guided by a program or remote control.

An **Android** is a specific type of Robot that is designed in the image and after the likeness of Human beings having the human form (Arm, Leg, Leg, Arm, Head, Torso) and human characteristics. Much like the Biblical story of God creating Humans "in his image and after his likeness", when Human beings create Robots in the Human image and after the Human likeness (which by the logic of the Bible then Android Robots would also be in the image and after the likeness of God because if A=B and B=C then also A=C), then it is called an Android coming from the Greek words "*andro-*" meaning "Human" and "-*eides*" meaning "form or shape". Based on the etymology of the word "Android", Human beings are "Androids" because all Human beings have "**Human form and shape**".

However, what happens when the "Human form and shape" is altered by either addition or subtraction of one of the fundamental component parts of the "Human form and shape"? If a person looses an arm and/or a leg, are they still in the "Human form"? Does the Human become any less-Human by losing a limb? Does the Human become any "more" Human by enhancing their "Human form"? At the nexus of these ideas is the third part of the triad of robotics called **Cybernetics** which are used to create "**Cybernetic Organisms**" or "**Cyborgs**" for short.

The word "**Cybernetics**" comes from a French word meaning "**the art of governing, establishing law, order, and control**" and is used in the fields of **Engineering** and **Computer science** as a term to refer to systems, structures, and devices used to regulate and control something. By definition, everything in Nature is "Cybernetic" because everything in Nature follows some governing regulations such as the "**laws of Nature**" or the "**laws of Physics**". When **man-made devices** are combined with **biological organisms**, then the being is considered a **Cybernetic Organism** or **Cyborg**. Considering this definition of "Cyborgs", then we must recognize the fact that people who live in Governments with man-made rules and laws are "Cyborgs" by definition. Also, considering that the word "Robot" actually means "slave" and the word "cybernetics" means "the art of Governing or creating laws", then we must recognize that people can become programmed and become Robot Slaves by the laws created by "Cybernetics" or Government. That's why you can observe people's mannerisms, body language, and speech pattern and determine what type of "program" is running on their "operating system" or Brain. Although most people are familiar with Cyborgs from Science-fiction movies and literature, Cyborgs commonly occur in "real life". Examples of "real life" Cyborgs are people who have received **Artificial limbs** or **prosthetic devices**. A Prosthetic is a device that replaces a missing body part. Prosthetic devices may or may not be electrical, mechanical, or include computer software. Examples of Prosthetic devices include artificial limbs like prosthetic arms, prosthetic hands, prosthetic legs, hearing aids, cardiac pacemakers, dentures, eyeglasses, walking canes, crutches, and wheelchairs. A **South African Cyborg Sprinter** named **Oscar Pistorius** competed in the **Summer 2012 Olympic Games**. Oscar Pistorius had both legs amputated below the knee and replaced with prosthetic devices called "**cheetah blades**". The **clothing we wear** is also a "Prosthetic" Cybernetic device because it is used to replace and/or enhance the outer covering of Human beings and also used to "control" and "regulate" our body temperature in an environment. To some extent, our house, cars, and cell phones can also be considered "Prosthetic" Cybernetic devices. **Project Glass** by Google is the development of another form of "real life" Cybernetic device which will be **"Augmented Reality"** eyeglasses which can scan bar codes and QR codes to obtain information and interact with the internet in real time while still observing "reality". Also, another "real life" story of Cyborgs occurred when the company "Bionic Vision" of Australia reported on August 30, 2012 that they had successfully implanted the world's first **robotic bionic eye** prototype in a woman with hereditary sight loss.

The earliest evidence of Prosthetic devices and Cyborgs can be found in Africa. A 3000 year old mummy of a woman who had her toe amputated and replaced with a **dark brown** wooden prosthetic was found in the Ancient Egyptian city of Thebes.

Above: African Cyborg Prosthetic toe from Ancient Egypt

Cyborgs represent the unification of dualities, **part man** and **part machine**. The etymology of the word "**Machine**" shares the same origin as the word "**Magic**". In Ancient Egypt, "Magic" was called **HEKA** (meaning to activate the **KA**) and is more closely likened to the modern concept of **Engineering**. The word "Man" shares the same etymological origin as the word "Mind". Therefore, the unification of the "part man-part machine" dichotomy found in Cyborgs can also be considered the unification of the "Mind" and "Engineering". The Unification of dual principles is a common theme found in traditional African cosmology and philosophy. The **Great Sphinx of Giza** with the **head of a Human** and the **body of a Lion** is a giant monument which also represents the **unification of dualities**. Thus, it is not surprising that the **Cyborg** named "**Darth Vader**" from the Science fiction movie **Star Wars** shares similarities in appearance to the Great Sphinx of Giza because both Darth Vader and the Sphinx represent **unification of dual principles**.

Darth Vader is a famous Cyborg from movies and appears to have been modeled after the image of the Great Sphinx

Another concept closely related to Cybernetic Organisms is the concept of "Artificial Life". **Artificial Life** is the study of how processes related to life can by modeled and simulated using Computers and Robotics. The **Reproductive** aspect of life can be modeled using Computer and Robotics. "Computer code which writes computer code", "Robots that build Robots", and "Machines that make machines" are all examples of simulating the **reproductive processes** of Life just like "Human beings making Human beings". **Artificial Intelligence** or **AI** is a subset of the field of Artificial Life which hones in on the "Intellect" or mental processes of life. One phenomenon that has been observed in Artificial Life simulations of Intelligence is that complex programs and software can be created which so closely match the way people think that the computer can actually "think" or "believe" it is a person. A non-profit organization called the "**Cyborg Foundation**" was founded in 2010 to meet the needs of the growing number of people who actually desire to become technologically enhanced by becoming a Cyborg. So now, what happens when the Android thinks it is a Human or a Human thinks it is an Android? The lines between what is "real" and what is "artificial" will become more and more blurred in the future as we come to the realization that the labels of "real" and "artificial" were a farce in the first place. Human beings feel that one quality that makes them unique is the fact that they are "**sentient**". However, the word "**sentient**" comes from a Latin word meaning "**to feel, to sense, to perceive**" and also shares etymological origin with the words sentinel, sentence, and syntax. Therefore, if a Human being looses their sense of sight, hearing, or ability to feel and perceive then they are no longer "sentient". Conversely, if a Robot or Android can be built with Sensors able to perceive sensations equal to or greater than Human beings, then the Robot or Android is also "sentient". Furthermore, Robotics can be combined with or added to Human beings who have lost their ability to "perceive" restoring their "sentient" qualities as Cyborgs. Just as cybernetic devices known as "clothing" enabled Humans to live on parts of the Planet in which they would not normally be able to survive, cybernetic devices like "**Space suits**" and "**scuba gear**" enable Human beings to travel into environments that would normally be inhospitable like **outer space** or **under water**. This realization that Humans are naturally limited to where they live and the "reality" they experience, and can only travel outside of the boundary of limitation by using cybernetics, shows the similarities between Human beings on Earth to **video game characters** in a **Virtual Reality** or simulation.

0111 - VIRTUAL REALITY IN AFRICA

Reality is defined in Philosophy as the state of things which actually exist. The study of existence and reality is called **Ontology**. Through Ontological studies, we learn that "Reality' is dichotomous composed of two principle parts which can be labeled as "**Abstract** or **Concrete**", "**Rational** or **Empirical**", "**Subjective** or **Objective**", "**Speculative** or **Operative**", or "**Hypothetic** or **Pragmatic**" respectively. To suggest that something "does not exist" is a logical fallacy because the mere moment that you are able to **talk about something**, or the mere moment you are able to **think about something**, it exists! Something may not exist empirically, or in the "real world", but it does exist at least as a concept, or rationally or abstractly, that is to say, IN THE MIND. This idea is the manifestation of the Ancient African concept associated with the deity **PTAH** who was said to have "created the Universe" (created the Virtual Reality) when it is said that "things" can be THOUGHT OF and SPOKEN into existence through "**creative speech**". The Muslims took this concept related to Ptah and later associated with their deity **Allah** were they say things were created into existence via the command "**Kuwn fiya Kuwn**" meaning "**be and it is**". By virtue of the fact that you are able to talk about something is evidence of the fact that the thing Exists.

Human beings exercise their abilities as creators, extensions of "The Creator", by turning concepts, ideas, and abstract things into manifested Realities through the process of **Creativity**. At the point where a Concept actually exists in Reality, its Ontological status has moved from the abstract to the concrete, but prior to the point of "**crossing over**", it was merely abstract. In order to bring an abstract concept into a concrete reality, it has to adhere to the laws of that reality; otherwise it will remain an abstraction. An abstract concept can be brought into reality and made to "seem" real by **speaking it into existence** or **talking about it** and **describing it** and the only limitations on bringing the abstraction into reality would be the ability of the language that is being spoken or written in to fully describe the concept. So the speaking or articulating of abstract ideas and concepts was the first **"Virtual Reality"**. Later when people began to write and draw, "Virtual Reality" was expressed in **written form** and in **picture form**, and the only limit on the expression of the abstraction was the ability to find a way to symbolically depict it with **logos**. In Africa, the stories told by **Griots** and the **fables** written down by **Scribes** are forms of Virtual Reality. In fact, the etymology of the word **"fable"** and the word **"simulation"** both come from words meaning **"false"**. All literary works of fiction, comic books, and even your ancient Mythologies like the Egyptian Mystery system, **Neteru**, **Annunaqi**, **Kachina**, **Nommos**, **Angels**, etc and their associated stories are all forms of "Virtual Reality" in so far as they exist as creations which are metaphors and similes used to symbolically simulate something and do not, and in most cases could not exist in the laws of this reality. In Ancient Egypt, a famous **"Virtual Reality"** or **"Simulation"** which was repetitiously run over time was the story of the **"afterlife"** or the **Duat** which first was relayed from "mouth-to-ear" as a story part of an oral tradition, and then later appeared as Hieroglyphics inscribed inside Pyramids called the **"Pyramid Text"**, then later appeared inscribed inside sarcophaguses called the **"Coffin Text"**, and even later was written down by scribes to create what is now known as **"The Book of the Dead"** or **"The Book of Coming Forth by Day"**. With the story of **"The Book of Coming Forth by Day"** inscribed in pictures and writing on the Pyramid walls, the Pyramid served as a Giant Virtual Reality simulator which enabled people who entered the Pyramid to experience the details of the story in vivid detail.

The book entitled "**The Science of Sciences and the Science in Sciences**" by **African Creation Energy** describes how the two dichotomies of reality were symbolized by "The **Moon and the Sun**" or "**Death and Life**" in Ancient African culture and philosophy. Comprehending the **dichotomy of Reality** and the African symbolism of "**the Moon and the Sun**", then the question must be asked: At dusk and at dawn, during the **twilight** hours, what is the symbolic significance of the Natural phenomenon which occurs when the Sun and the Moon can be seen in the sky at the same time? The answer to this question is "**Virtual Reality**".

We must comprehend that the many sensations, feelings, and interactions we experience in the "Real" world are all translated to our brains as electrical signals; these facts have led to some scientist calling our reality the "**Holographic Universe**". This is why we can "Dream" while we are sleeping and have an experience which seems "**as Real as**" or "**as believable as**" any "Real" experience we would have in "Reality" while we are awake. The definition of "Virtual Reality" is any "Reality" or experience which is "as real as" or "as believable as" Real reality. There is a term called "**Lucid Dreaming**" which refers to a state where the dreamer is conscious of the fact that he or she is dreaming and is able to control and manipulate the events of the dream and the imaginary environment of the dream at will. It is believed that individuals who are able to take control of their dreams in the state of "lucid dreaming" are better able to take control of their Reality and Life to shape it, steer it, and guide it to gain desired results. Likewise, a "**Lucid**

Virtual Reality" experience should empower individuals to be able to learn how to take control of their life in the "real world".

Virtual Reality builds on the principle that our thoughts and our experiences are the result of electrical signals in our brains, and therefore "Reality" can be simulated using a variety of methods to deliver the electrical signals of an experience to our brain. Virtual Reality is the term used when the method of simulating reality is generated by computers; however reality can be simulated through a multitude of other sources including stories, books, and paintings. The invention of the computer is just another technology that enabled the means to express and create new abstract things. So as Technology progresses, the ability of Human Beings to make their abstractions seem more concrete gets better and better, and the abstract idea is able to seem more and more real. A black and white drawing can seem more real than a written description, and a color drawing can seem more real than a black and white drawing, and a **Pop-up book** can seem more real than a regular flat illustrated book, and an animated drawing can seem more real than a still drawing, and a **Hologram** or **3D** animation can seem more real than a 2D animation.

Anything that is created is a relative "Virtual Reality" since it was brought into the empirical world from the mental realm by way of the human creation process and creativity. When an environment is created either by way of rules and laws, and/or by way of tools and structures, then you are creating a virtual reality. So when an environment is created on top of, or within, an existing environment, a "Virtual Reality" is simultaneously created, i.e. a reality that is similar, but not exactly like the reality that it was built upon. Therefore your house is a "Virtual Reality," and your school is a "Virtual Reality". One of the characteristics of a "Virtual Reality" is that there are rules and laws that pertain to the created reality that do not necessarily pertain to the reality that it was created in.

Another characteristic of a Virtual Reality is that it enables the Mind to be taken to "another world" so to speak. Movies, Books, Video Games, Pictures, Stories, Structures, and Buildings, all are the result of Brain Concentration and the Creation of things that can take the mind to another "Reality" or a Virtual Reality. In the future, the **Video Game industry** will likely merge with the film, movies, and **Cinema industry** to start creating a new form of **interactive Virtual Reality video-game-movies**. Therefore, knowledge of Computer programming and computer graphic design will be necessary for African filmographers and African movie producers in the near future.

The "Virtual Reality" Phenomenon can also be observed in **Quantum Physics** and **Special Relativity**. The word "Physics" is a Greek word meaning Nature, and the laws of classical Physics or **"laws of Nature"** that seem to perfectly describe our world or relative reality, do not work and do not describe the Laws of the Sub-Atomic or Quantum World. Also, the laws that seem to describe things on the Astronomical level in Space like Planets, Solar Systems, Galaxies, and Universes, do not necessarily apply to our level or the sub-atomic level. Therefore, we observe that there is **Relativity in Reality** and these different levels are all different Realities, all which are built on the **Ultimate Reality** which is the **Quantum world** of the smallest Particles of Matter called **"Fundamental Particles"** in Physics. Each of these "relative realities" can be considered **"Alternate Realities"** or **"Parallel Universes"** to each other with the collection of all "Parallel Universes" making up what would be called the **"Multiverse"** to acknowledge the multitude of various "realities" which exist within the entire Universe proper.

Just like a Virtual Reality simulation in a Computer is composed of **digital bits of information** which can be considered **"Fundamental Computer Particles"**, the Reality of the Universe is composed of the smallest Particles of Matter called **Fundamental Particles of Nature.** Since we comprehend that the digital bits which make up the "fundamental computer particles" are actually Quantum electron and photon particles, then we know that the answer to the question **"Where is Cyberspace"** is **"In the Quantum subatomic world of Electrons and Photons"**. Since the environment or "reality" here on Earth is not the same environment or "reality" that is consistent throughout the Universe, then we must acknowledge that relative to the reality of the Universe, Earth is a "Virtual Reality". Just as **video game characters** are dependent upon the "reality" of the video game for their existence, Human beings are dependent upon the "reality" of Earth for our existence. If the **video game Virtual Reality** is turned off, then the video game characters cease to exist. If the environment which makes up the "reality" of Earth ever is "turned off" or ceases to exist, then the Human being will cease to exist if the Human being does not learn how to adapt to new and different realities. This is why the preservation of Nature is so important in traditional African culture, because African philosophers realized that Human Beings are metaphorically **"Video Game characters"** living in the **"Virtual Reality"** of Earth's environment and this sentiment is expressed in a proverb from the **Pygmy** tribes in South and **Central Africa** which states **"Our society will die if the forest dies"**.

Time is a measurement of Existence. Therefore, in a discussion about "Virtual Reality" or **"Virtual Existence"** and "Relative Reality" or **"Relative Existence"**, we must also consider the concept of **"Virtual Time"** or **"Relative Time"**. Through **statistical predictive modeling** it is possible, with the collection of sufficient data, to use the computer to create Virtual Reality simulations which **reconstruct the past** or **predict possible futures**. Therefore, it is likely that the realization of a **"Time Machine"** will most likely be a **Virtual Reality simulation** which allows the user to travel forward or backward in time, or even experience an alternative "present" if certain prior decisions the user made in the past were different.

The Ancient Africans in Egypt personified the concepts related to **Time Travel** in a deity named **Khonsu** who was depicted as a falcon with **two heads looking in two directions**

Khonsu and the Ancient African Secret of Time Travel through Virtual Reality

(past and future). The name "Khonsu" meant **"traveler"** and was associated with the **"traveling"** of the **Moon across the sky** but could also be used to describe the "travelling" of the **Sun** across the sky. Khonsu was thought to be an **"old man"** in the **evening** and a **child** in the morning, which meant as the moon traveled through the sky at night, **Khonsu traveled "backwards in time"** or "backwards in age" like the main character in the year 1922 book and year 2008 movie entitled

"The Curious Case of Benjamin Button". Therefore, Virtual Reality is an operative expression of the African philosophical doctrine of the **"Sun and the Moon"** and the deity **Khonsu and the Ancient African Secret of Time travel**.

While Virtual Reality is a complete and total simulation of Reality, **"Augmented Reality"** is a partial simulation of reality where computer generated sensory information is added to the physical "real-world" environment. The most common Augmented Reality sensor inputs are sights and sounds. Augmented Reality sounds are created by using computer generated sound through devices worn on the ears. Augmented Reality images are created electronically using computer generated images in eyeglasses or in contact lens or by projecting **Holographic** images into the "real world".

Holograms occur naturally in Nature. **Rainbows** and **aurora borealis** (or the northern lights) are examples of **naturally occurring Holograms**. In the city of Timbuktu in Mali West Africa, a phenomenon known as a **"Sun Dog"**,

Above: a Parhelion Naturally occurring Hologram of 3 Suns

Parhelion, or **"22° halo"** occurs when by ice crystals in the air act as prisms which bend sunlight to create an optical illusion of two **Holographic suns** sitting on either side of a **central sun** giving the appearance of **3 suns in the sky**. The phenomenon can also appear as a large ring or halo around the sun.

Prior to the invention of the Computer and electronic devices, the usage and consumption of **drugs** and/or **alcoholic beverages** which produced **Psychedelic** effects were used by people to "**Augment**" and **escape reality**. Likewise, it is possible to create "**digital drugs**" which allow people to escape reality through the use of **Augmented Reality** and **Virtual Reality computer simulations**. Just like drugs, alcohol, and other **escapism mechanisms** like "fantasy lands", theme parks, and religion which stifles people's productivity and enables people to **run away from their responsibilities** and not deal with the real circumstances they are faced with in life, Virtual Reality and Augmented Reality products will be the "drugs" and alcohol of the future which will have to be regulated and controlled. "**Electronic Addition**" can already be observed through the use of social networking websites like **Facebook** and **Myspace**, or through the use of massive multiplayer role-playing games or simulations like **World of Warcraft**, **Second Life**, **The Sims**, and **Technosphere**. So while conspiracy Theorists were concerned about the "Illuminati" Boogie Man **putting a computer chip inside them**, they completely ignored the other side of equation of "**putting people inside of the computer chip**". What is meant by "putting people inside of the computer chip" is when people voluntarily put all their personal information and pictures into a website on the computer to "**profile themselves**", and even place their mind and sensory organs "inside" of the computer by way of Virtual Reality and Augmented Reality simulations; Literally putting themselves inside the "Chip" or inside the Computer. In the future, the lines between Reality, Augmented Reality, and Virtual Reality will become blurred. The blurring of the lines of reality is explored in movies like the **Matrix trilogy**, **Terminator Salvation**, **The 13th floor**, **Surrogates**, and **Avatar**.

Since it is possible for a creation to not only become like the creator, but to also to surpass the creator, then it is also possible for a created "Reality" to be relatively "better" than the actual reality it was built upon. There is a large segment of the Human population that prefers fantasy and fiction over reality and truth. People who have a propensity to believe and accept **fantasy and fiction** will likely become **addicts** and **consumers** of Virtual Reality and Augmented Reality computer simulations in the future. People who have a propensity to only accept **facts and the truth** will likely become the **creators** and **producers** of Virtual Reality and Augmented Reality computer simulations in the future. This dichotomy which exists amongst people is why in the **Matrix** movie, some people voluntarily choose to take the **"Blue Pill"** which was a metaphorical way of choosing to accept fantasy, fiction, and simulations over truth and reality. So you have to decide, are you a **Programmer** or are you a **Robot**? Are you a **Creator** or are you a **Consumer**? Are you a **Dealer** or are you a **User**? Are you a **Victim of Circumstance** or are you a **Creator of Circumstance**? Do you **use the computer** or do you **let the computer use you**? Do you create your environment and control your Reality, or does your environment effect and control you?

Creating "Virtual Realities" is a way to "think outside the box". **"Thinking Outside the Box"** means thinking outside of your relative reality or created reality because the **"box"** or the **"square"** is related to the Mind and Created things. **Free will** and **choice** is really an **illusion** because you only have the will and freedom to choose from choices and options made available to you if you exist in a reality created by an intelligence other than your own. However, as the creator of a "Reality" you make options available to others, and you know the outcome regardless of which option participants in your reality select. Given the choice of one option, you can take it or leave it. Given the choice of two options you can have option 1, have option 2, have both option 1 and option 2, or have none (have neither option 1 nor option 2). Given the choice of three options, the **combinations** and **permutations** increase. Since your options are fixed in an existing reality, a plan can be made for each choice, and a path or program can be set based on the options made available to you. Even when you create or make something in an existing reality you have merely combined or destroyed existing options and therefore a path or program can be established for anything you "create" in an existing reality. Therefore, the options of your reality can serve as a **predetermined** or **predestined program** of your life with you as the "robot" or "video game character" serving your function and purpose which is to choose. Virtual Reality designers and Video game designers do not participate and play video games for the same reason that video game consumers play video games. Video game designers already know the outcome of every choice they make within the video game because they had to account for it in the design of the game. Video game designers play video games to ensure that the game is operating and functioning in the way that they desired in their design and if they encounter anything undesired they have the power and ability to change it. As mentioned earlier, the Ultimate Reality on which all other "realities" are built is the Fundamental Particles and Forces of Nature. If you have the ability to make changes to the Fundamental Particles and Forces of Nature, then you are likely the creator and designer of the Ultimate Reality. However, if you cannot make changes to the Fundamental Particles and Forces of Nature, then you already exist within a Virtual Reality. Therefore it is import not to worry about things not in your control but also be realistic about what things you can control.

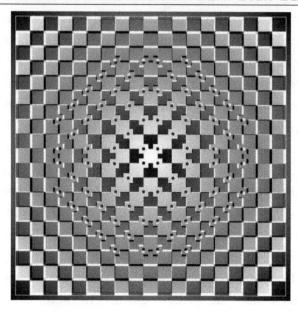

When you stare at the center of the picture above, if the sides and corners appear to move, then you are experiencing a "Virtual Reality" because the reality is that the picture is not moving. **Nothing and NOONE can tell you the truth**, people tell you statements, but **truth and falsehood** are both things you must determine for yourself using **your own logic and reason**. Can the Sun lie to you? Can Nature lie to you? Can your eyes and ears lie to you? Can you lie to yourself? **Optical Illusions** naturally occur in Nature. Holograms naturally occur in Nature. Virtual Reality naturally occurs in Nature. If you choose to accept something as true that is not actually true, that is your own fault. You experience life and you determine if your experiences are true or not. Stop placing responsibility outside of yourself. Making statements like "**You lied to me**" is a way for you to **escape the responsibility** of using your own logic and reason to determine if the statement that was initially made was true or false in the first place.

Truth and Reality are things

You should determine for Yourself!

1000 - Transhumanism in Africa

What does it mean to be Human? Some Humans would suggest that the very ability to even ponder the question about "**What it means to be Human**" is the definition and meaning of being a "**self-aware**" Human and part of the Human condition. In the Science of **Biology**, being "Human" is classified under the **genus Homo** coming from the word "*humus*" meaning "**ground, dirt**, or **earth**". The word "humus" is also phonetically similar to the Ancient Egyptian word "**Hu-mose**" meaning "*child of the creative force of will*". Modern Humans are a species of the genus Homo called "**Homo-Sapiens**" with the word "*sapiens*" coming from a word meaning "**wisdom** and **understanding**" with the root word "*sap*" meaning "**juice, fluid**, or **nectar**." The word "Homo-" in "Homo-Sapiens" comes from the Latin language and means "Human" and should not be confused with the phonetically similar word "**Homos**" from the Greek Language which means "*one and the same*" as in the word **Homogeneous**. The ironic thing about the definition of "Homo-sapiens" being "wisdom and understanding" is that one of the primary characteristics of the "**Human Condition**" is to be curious and ask questions such as "what is the meaning of life" and "what will happen after death." Perhaps rather than being called "Homo Sapiens", Humans should be called "Homo seeking sapiens" which would better indicate the curious nature of the Human who asks questions seeking "wisdom and understanding". One of the defining characteristics of the genus Homo is the **use of stone tools** or "**technology**" as an extension of one's self. The use of technology and the asking of **questions** are indications that part of being Human is the desire to go on a **quest** to be better than and go beyond their current state of being. If one of the main characteristics of Humans is to ask questions, then being Human means "**to not know**" or "**to be**

ignorant" which then justifies the need to ask questions so that Humans can seek and gain knowledge. The word "Homo-Sapiens" as the definition for Humans is an **oxymoron** like the word "**sophomore**" which means "**wise fool**" from the Greek words "*sophos-*" meaning "**wise**" and the Greek word "*-moros*" meaning "**foolish** and **dull**". The word "**Trans**" means "**to move or go beyond**", thus the word "**Transhuman**" indicates "**moving or going beyond the qualities which define being Human**". If one of the qualities which defines being Human is to ask questions and identify problems, then if someone has **answers to questions**, and **solutions to problems**, then that someone is indeed "**beyond Human**" or **Transhuman**. Problem solvers are Creators (**Khaliqians**), and Creators are Providers (**Rizqians**). The individuals who solve the problems of Humans, Create for Humans, and Provide for Humans have always been considered "beyond" Humans (Transhumans) and have historically been honored, revered, immortalized, and even deified by Humans. There are individuals reading this that want to worship, and there are individuals reading this that will be worshipped, and the individuals reading this that will be worshipped are the Transhumans.

In the Science of **Genetics**, **DNA** is the defining characteristic of being "Human". The basic elements that make up Human DNA are **Carbon, Hydrogen, Nitrogen**, and **Phosphorus**. Carbon, Hydrogen, Nitrogen, and Phosphorus are composed of **Protons, Neutrons**, and **Electrons**. Electrons are a fundamental Particle of Nature, but Protons and Neutrons are composed of **Quarks**. Therefore, **Electrons and Quarks** are the Fundamental Particles of Nature which are your "**DNA's DNA**", that is to say, the substances which your DNA is composed of, and which preceded your DNA. Electrons and Quarks are the "mothers and fathers" of DNA. Therefore, at your essence, Humans are electrons and quarks.

In Religion and Theology, being "Human" is the opposite of being divine. Divine beings are Immortal and not subject to death, whereas one of the defining qualities of being Human is to be **Mortal** and **die**. The word "**Mortal**" means "subject to death, deadly, and doomed to die" and comes from a Proto-Indo-European word "***mer-***" meaning "**to die**". This characteristic of mortality being one of the defining qualities of Human beings is expressed in the Judeo-Christian Bible in the book of **Genesis 3:22** where it states:

*"And the Lord God (**Yahweh Eloheem**) said, Behold, **the man is become as one of us**, to know good and evil: and now, lest he put forth his hand, and take also of the tree of life, and eat, and **live forever**"*

In the Judeo-Christian Bible, the "**Lord God**" (**Yahweh Eloheem**) evicted Human beings from the **Garden of Eden** because Human beings had disobeyed the instructions or program given to them, and the "Lord God" did not want Human beings to "live forever" with the ability to disobey. However, Genesis 3:22 of the Judeo-Christian Bible indicates that Transhumanism is a part of the Judeo-Christian Religions because the statement *"the man is become like one of us"* indicates that the Human being "**became**" or **Transcended** their original design to **become** like the "Lord God" (Yahweh Eloheem). **Transhumanism** can further be seen in the Christian religion in the concept of **Transubstantiation** in Catholicism where the "body and blood" of Christ is believed to have **changed into new substances** such as "bread and wine" to be eaten for communion. The philosophy of Transhumanism also appears in the Christian New Testament of the Bible in the book of **1 Corinthians 15:51-54, 58** where it states:

*"(51) Behold, I show you a mystery; we shall not all sleep, but we shall all be **TRANSFORMED**, (52) in a moment, in the twinkling of an eye, at the last trump: for the trumpet shall sound, and the **dead shall be raised** incorruptible, and **we shall be changed**. (53) For this corruptible must put on incorruption, and this **MORTAL must put on IMMORTALITY**. (54) So when this corruptible shall have put on incorruption, and this mortal shall have put on immortality, then shall be brought to pass the saying that is written, Death is swallowed up in victory... (58) Therefore, my beloved brethren, be ye steadfast, unmovable, always abounding in the **work** of the Lord, forasmuch as ye know that your **labor** is not in vain in the Lord."*

The philosophy of Transhumanism also appears in the Islamic religion. In the Islamic religion, there are two **Hadiths** or sayings of the **Islamic Prophet Muhammad** which allude to Transhumanism. One Hadith states that the Islamic Prophet Muhammad said that "**Black Seed** *is a* **cure to everything** *except* **old age** *and* **death**" (which suggests there must be a cure to old age and death). Another Hadith states that the Islamic Prophet Muhammad said *"There is no disease that Allah has created, that Allah did not also create a remedy for."* These saying of the Islamic Prophet Muhammad allude to Transhumanism because they suggest that there is a solution and a cure to "old age and death". Also, in **Sura Insan**, **chapter 76 verses 1 and 2** of the **Quran** where it states:

*"(1) Has there not been **over Man a long period of Time when he was nothing not worth mentioning**? (2) Surely we created Man from a drop of mingled semen, in order to try him: so we gave him hearing and sight."*

In the Islamic Religion, there was a long period of time when Humans did not exist, and there will be a long period of time when Humans will not exist, therefore in the overall scheme of time, the existence of Human beings is **"not even worth mentioning"**. Modern historians credit the Greek philosopher **Heraclitus** with saying **"The only constant in the Universe is change"**. Heraclitus is also credited with establishing the term **"Logos"**. However, the book **"Stolen Legacy"** by G.M. James describes how Heraclitus derived his philosophy of the logos and his philosophy of change being central to the universe from the **Memphite Theology** of **Ptah**. Therefore, the Islamic idea that in the overall existence of the Universe, the period of time in which Human beings exist is "not even worth mentioning" is due to the reality that **the only constant in the Universe is change (Trans)**, and the state of being Human is just a temporary condition in the many Thermodynamic changes that will occur over time during the life of the Universe.

Therefore, Human beings are not as important or special as we would like to think. There have been species that have existed on this planet that are now extinct. There have been other species of genus Homo which have existed on this planet like the **Neanderthals** which died out about **24,000 years ago** and are now extinct, and there are certain Human genetics that are now on this

The Evolutionary March of Mankind

planet which are **Endangered Species** on the verge of extinction. The change over time with the rise and fall of various species can be seen in the famous graphic of the **"evolution of mankind"**.

The only constant in the Universe is **change (Trans)**, and anything resistant to adapt to the changes of the Universe eventually meets their end.   Therefore, Transhumanism, or the philosophy of "Trans" is a Universal philosophy in Nature.   The etymology of the word **"Trans"** comes from the Latin words *"trare-"* or *"tere-"* meaning **"to cross or to go through"**.   The word "Trans" also comes from the Avestan word *taro* meaning **"through or beyond"** which is also related to the **Islamic Sufi** principle *Turuq* or *Tariqa* **(Paa Taraq)** which means **"way, path**, or **method"** and the Arabic word **"taraka"** meaning **"to leave behind, abandon, or omit"**.   The philosophy of Transhumanism also appears in the doctrine of Far East religions such as Buddhism and Hinduism in concepts   such   as   **Transcendence**   and   **Transcendental Meditation**.   All of these words related to the word **"Trans"** share the same phonetics and meaning with the Ancient African word **Ptah**.

Transhumanism in the contemporary sense of the word is defined as a movement which aims to transform the human condition by way of Science and Technology.   Transhumanist thinkers speculate about the potential outcomes that may arise in the future for Humanity using the term **"Post Humanity"**.   Recall that the word "Trans" means "to move", and therefore movement can be for better or for worse.   Transhumanism ideas which seek to improve or transform the human condition for the better are abbreviated by the symbol **H+**.   Transhumanism H+ is also called by the term **Homo-Superior** which is used to refer to the point where Humans merge with Machines.   However, if we consider that one of the characteristics of genus Homo is the use of tools and technology, we could argue that this state of "merging with machines" has already occurred or was on the verge of occurring since the advent of genus Homo.   Transhumanism ideas which seek to transform the human condition for the worse are abbreviated by the symbol **H-**.   While H+ is to become greater than Human, H- is to become less than Human.   Science and technology which creates alcohol and drugs like Crack Cocaine and Ecstasy which alter the mind, or the process of genetically altering foods for the worse which in turn alters the body for the worse are all examples of H- Transhumanist ideas and practices. Authors and speakers including **Ben Ammi**, **Dr. Malachi Z. York**, **Dr. Delbert Blair**, **Dr. Phil Valentine**, **Amun-Re Sen Atum-Re**, and **Sadiki Bakari** have all written books and produced videos geared towards Africans and people of African descent discussing the potential dangers that could arise from the misuse of science and technology. The book "**Dehumanization**" by Ben Ammi and the three-part video series entitled **Transhumanism I**, **II**, and **III** by Sadiki Bakari are excellent resources which detail the possible peril that could arise from negative Transhumanism ideologies.

Suspended between the Transhumanism concepts of H- and H+ would be the third point of "**H**" which would represent "**no change**" or "**remaining Human**". Recall that **the only constant in the Universe is change**. A desire to remain the same forever with no growth or no development is **stagnation** and **complacency** and synonymous to **death**. Anything that is not changing and not growing is "dead", and even things that are "dead" are subject to the laws of change in the Universe. There is **Positive Growth** (Construction, H+) in Nature, and there is **Negative Growth** (Destruction, H-) in Nature. Since Transhumanism aims to transform the Human condition, there are a variety of ways in which **Human Transcendence** can occur including **Biological Transhumanism** and **Technological Transhumanism**.

Biological Transhumanism is the transformation of the Human condition using Biological methods. An example of Biological Transhumanism is **Eugenics**. The goal of Eugenics is to improve the genetics of a population. One of the ways the genetics of a population is improved is by **Selective breeding**. The **Eugenics** process of **selective breeding** to **transform** a human population was and is widely used in Traditional African culture. In traditional African culture, only certain tribes are allowed to marry and breed with each other. African tribes which have traditionally been **Blacksmiths** or **Priests** are expected to only marry within the tribe as a way to strengthen the genetics required to perform the duties of a Blacksmith or a Priest. In Ancient Nile Valley culture, certain **genetic bloodlines** were responsible for being Pharaoh, Ruler, Vizier, or Scribes and the Eugenic process of selective breeding was practiced as a method to keep the bloodline pure and untainted to further strengthen the leadership or intellectual qualities generationally over time. Even a parent encouraging their child by saying "**Don't be like me, be better than me**" is a form of Transhumanism because the parent aspires for the child to **Transcend** the realities, conditions, and circumstances that the parent experienced.

In modern times, the term Eugenics has taken on a negative connotation because of the activities of **Adolf Hitler**, the **Nazis**, and people who prescribe to **racist "Social Darwinism"** ideologies. Just like there is **constructive** and **destructive** Transhumanism, there is also constructive and destructive Eugenics. Selective Breeding is a constructive method of Eugenics where the person you choose to have a child with is selected based on **positive qualities** such as **Intelligence**, **morality**, **character**, and **family history** which will ultimately breed a more intelligent and more moral child. However, Selective Breeding can also be corrupted when the criteria used to select a breeding partner is based on **superficial qualities** such as hair length, the size of a woman's breast, or the size of a woman's hips and butt, or the size of a man's muscles. Selective breeding based on superficial qualities ultimately leads to breeding superficial children. Moreover, with the use of **enhancement technologies** called **makeup**, **plastic surgery**, **hair weave** and **hair extensions**, the superficial qualities of "**beauty**" are able to be synthesized. Women are able to increase the size of their breast using **breast implants** and **breast augmentation**. Women are able to increase the size of their hips and increase the size of their buttocks with "**butt shots**". Men and Women are able to decrease the size of their waste line with **liposuction**. Men are able to increase the size of their muscles using **steroids**, **testosterone supplements**, and **weightlifting**. All of these are examples of various forms of "**enhancement technologies**" which ultimately create **real-life Cyborgs** whose purpose is to alter who you would **naturally select** to breed with based on the **appearance criteria** of **selective breeding**.

However, regardless of the criteria used for **selective breeding**, the result of the action is still the creation of a new life and in general is still not as destructive as the Eugenics process of Genocide which produces **death**. **Genocide** is defined as the deliberate and systematic reduction of a targeted genetic group. Genocidal activities include using **birth control**, **abortion**, **sterilization** of people with certain genetics so that they cannot reproduce and also the outright **killing** and **murder** of people with certain genetics significantly reducing or eliminating that gene from the planet. It is the negative actions of Genocide which have given Eugenics and ultimately Transhumanism a negative connotation. Genocidal activities have historically been carried out by people who prescribe to **racist** doctrines of **superiority** or people who prescribe to "**Social Darwinism**" and "**Survival of the Fittest**" philosophies. However, the built in logical fallacy of people who prescribe to "**Superiority Theories**" or "**Survival of the Fittest**" theories and carry out Genocidal actions against other groups of people is that **if you are indeed Superior and if you are indeed the "fittest" then you should not have to kill or eliminate your competition.** The fact that someone feels the need to eliminate or hinder their competition is a testimony to the fact that they feel **inferior** and they feel that they can in fact be beaten by their competition. If you were truly Superior, then you would not have to hinder the inferior being in any way. If you were in a competition where you were one of the four best sprinters in the world, in order to prove you are the best sprinter in the world, you would want the other 3 sprinters to arrive at the race in their top form so that when you beat them, then it will be obvious to everyone that you are the most superior sprinter racing. However, if you feel the need to drug your competitors, or eliminate your competitors, or trip your competitors during the race so that they cannot run at their peak performance, then you are acknowledging through your actions that you feel that the other runners are indeed superior to you and can in fact beat you; this is the logical fallacy which exists in the mind of people who prescribe to racist doctrines and perform genocidal activities against people of other races. **Racism** is defined as a **prejudice based upon race**, and therefore anybody of any race can be a racist. To suggest that only one group of people on the planet are capable of being racist is to simultaneously suggest that that group of people is "**Superior**" or "**Supreme**", i.e. **able to do something that no one else on the planet can do**. In life, it may be unavoidable to prejudge someone based on their race or any other criteria, however when evidence is experienced which contradicts your prejudgment, an intelligent person should abandon their **faulty assumption.**

The philosophy of Biological Transhumanism can be found in Ancient Nile Valley culture in Africa. In Ancient Egypt, many of the deities or "gods" were depicted as **Anthropomorphisms combining Human characteristics and features with the characteristics and features of animals.** Since the gods of Ancient Egypt were considered greater than human, and since the gods were depicted as part human and part animal, then the Ancient Egyptian deities can be considered the **origin of Biological H+ Transhumanism.** Although the Ancient Egyptian Deities were considered "greater than Human", it was believed that Human beings could personify, manifest, channel, and take on certain characteristics of the gods and thereby Transcend being Human (Transhuman) to become diving or "god-like". The names of some Ancient Egyptian deities which exhibit examples of Biological Transhumanism qualities are below:

- **Anupu** – head of a dog, body of a man
- **Ba bird** – head of a human, body of a bird
- **Bast** – head of a cat, body of a woman
- **Khnum** – head of a Ram, body of a man
- **Heru** – head of a falcon, body of a man
- **RA** – head of a hawk, body of a man
- **Sekhmet** – head of a Lioness, body of a woman
- **Sobek** – head of an Crocodile, body of a man
- **Tehuti** – head of an Ibis bird, body of a man
- **Tefnut** – head of a Lioness, body of a woman

The **Har-em-akhet** (Great Sphinx of Giza) with the head of a Human and the body of a Lion stands as a giant monument in Africa representing the philosophy of Biological H+ Transhumanism.

Head of a Human, Body of a Lion
A Monument to Transhumanism in Africa

The prevalence of the Half-human-half-animal Biological H+ Transhumanism paradigms which existed in Ancient Nile Valley culture is a manifestation of the Ancient Egyptian philosophy of **"Smai Tawi"** which referred to **"uniting the two lands"** but also dealt with the **unification of dual principles** in nature. The philosophy of Biological H+ Transhumanism can also be found in other African cosmologies including the **Dogon** story of the **Nommons** and Humans mating with reptilian/fish-like "Gods", or the **Akan** story of **Anansi**, the **spider who was also a man**. This combination of genetics to **biologically go beyond being Human** was depicted in the year 2009 movie entitled **"Splice"** where the genetics of multiple creatures were combined to create a Transhuman creature.

However, although Human beings differentiate animals and species, the Biological concept of **Homology** is the process of finding similarities between species. In Biological Homology, any traits or characteristics that are similar between species are considered to be analogous or Homologous, and therefore those species are theorized to have been derived from a common ancestor. Therefore, when we observe that all **vertebrate creatures** in Nature have the same basic form (Arm, Leg, Leg, Arm, Head, Torso – or some homologous variation thereof), then we recognize the possibility that all vertebrate creatures share a common ancestor which means that Humans are here today because we **Transformed** from one thing to another in the past. The philosophy of Transhumanism considers what might be the next change or transformation after Human. In considering how we have changed in the past, we can make predictions about how we may biologically change in the future. Any appendage on the Human body that is not completely controllable is subject to disintegrate over time, therefore future Human beings will likely lose their ring finger, 1 or 2 toes, ears, and nose. If you don't use it, you lose it. **Vestigial appendages** or **appendages without bones are subject to disintegrate over time**, therefore future Human females may not have breast and future Human males may not have a phallus. As Human beings become more intelligent, the Human brain and skull will likely get larger, and therefore future Humans will likely have bigger heads. The creature described which could result from Biological H+ Transhumanism bears a striking resemblance to what modern Humans have called the "**Grey Aliens**". One Biological Transhumanist theory is that these Grey Aliens that people have been seeing are in fact future Trans-humans who have travelled back in time in space ships to observe modern Humans and influence the course of Human events. Although this theory sounds far-fetched, we must recognize that every depiction of the Grey Aliens is that of a vertebrate creature with the Arm-Leg-Leg-Arm-Head-Torso form and therefore by Biological Homology, if Grey Aliens are real and truly exist, then they share a common ancestor with Human beings. Furthermore, the "**flying saucer**" crafts that people associate with the Grey Aliens bear striking resemblance to saucer-shaped crafts based on "**Helicopter**" technology built here on Earth. Thus, it may be possible that just like the Grey Alien is the future of Humanity, the "flying saucer" is the future form of travel developed by future Humans. Only time will tell what Biological **Transformations** will occur in the future.

Biological Transhumanism

The term "Transhumanism" is most commonly used in contemporary settings to refer to **"Technological Transhumanism"**. One of the meanings of the word **"Technology"** is **"to apply knowledge"** and therefore Technology not only refers to gadgets, gizmos, and electronics, but Technology also refers to anything created by "applying knowledge" and thus would also include social structures, religions, agricultural systems, economic systems, and governments. Between Biological Transhumanism and Technological Transhumanism, there is a **"Bio-Techno-Transhumanism"** idea that rather than having Human beings directly merge with machines, technology can be used to alter Human DNA with Biological Transhumanism being stimulated by Technology. Prescribers to **Technophobic Conspiracy theories** of a **dystopian future** suggest that the way Technology will alter Human DNA is by the government secretly placing **nano-robotics** into **vaccines** so that when the vaccine is administered, the nano-robotics will go to work rearranging your DNA. While the Technophobic conspiracy theory of nano-robots rearranging Human DNA could happen, the most likely way Technology will alter Human DNA is the same way Technology has been altering Human DNA for millennia, and that is through the Human's direct use of Technology. It is important to acknowledge that a House, Clothing, Shoes, and a car are all technology. Since our Ancient ancestors could survive outside without a house and without clothing and without shoes, but modern day Humans cannot, then this is evidence that Human DNA has already been altered through the utilization of technology. Moreover, since DNA is composed of something (electrons and quarks), this is evidence that DNA is not fundamental and thus has an origin. Since DNA has an origin, this means that DNA was made or created. Therefore, since DNA has an origin and was made or created, then **DNA itself is a form of technology**.

Beyond Technology altering, influencing, or changing Human DNA through use, there is the Technological Transhumanism idea that Human beings will go beyond being Human by directly merging with computers, technology, and machines. To what degree this merger with computers and technology will take place varies, but what is consistent is that Technological Transhumanism will result in a **New Being**. **Moore's Law**, **Ubiquitous Computing**, and **Cybernetics** are the three fronts which are expected to converge which will lead to Technological Transhumanism. As mentioned in the earlier chapters of this book, Moore's Law describes the exponential growth of computing capabilities. Ubiquitous Computing is basically when computers and technology will become so pervasive where the technology will be seemingly everywhere: in clothing, shoes, hats, eyeglasses, walls, cars, etc. With the growth of Ubiquitous computing, clothing will literally become a cybernetic device. Based on Moore's Law, when the computing capabilities surpass human mental abilities and Humans are already "partially" cyborgs with computers in their clothing, then people may begin mentally merging with the computers and technology. Technological Transhumanism suggests that the merger of Humans with computers will result in the creation of a New Being with superior intelligence. This point where Humans merge with computers has been termed the **Technological Singularity**, and it is the point in which **Futurologist** who make **statistical predictions into the future** foresee so much change that accurate predictions cannot be made, and thus any potential future is possible. The Technological Singularity and Technological Transhumanism is expected to occur sometime between the years **2033** and **2050**, and is expected to be the most significant event in the history of Humanity. Some people fear Technological Transhumanism. People fear what they do not comprehend, and what is funny is that most people do not even comprehend how their current Human body works but they are not afraid of it, and so if they do not comprehend how a cybernetic body works, they should not be afraid of it either.

Technological Transhumanism

What people are truly afraid of is losing their ability to "**choose**", and losing their ability to be "**unique**", and losing their ability to "**rebel**" and "**do what they want**". However, as discussed in previous chapters of this book, "choice" is an illusion. Even when you choose to rebel, you are making a choice that has already been accounted for. The fact that even the choice to "rebel" can be programmed and accounted for was depicted in the **Matrix** movie with the fact that even if you "woke up" and rejected the Matrix computer simulation and joined the Human city in "Zion", this was still a system of control or option that was provided to you by the designer of the Matrix and therefore you were still being controlled. Thus, the only way to not be controlled is to be the creator of the systems which control others. In religious Theology, the creator of the systems of control is called "God". In Ancient African culture in Egypt, the **Pharaoh** could **Transcend** being Human to be considered a **God** to his subjects. Moreover, ceremonies like the "Opening of the Mouth Ceremony" discussed in earlier chapters of this book, and **mummification** served as technologies which enabled the Pharaoh to **Transcend death** and live forever.

One of the expected consequences of Technological Transhumanism is the ability to have an **extended lifespan** or potentially to **never die** and **live forever** or to **resurrect** people who have died. The Technological Transhumanism idea of eternal life or resurrection after death by way of technology is the manifestation and practical application of the Ancient African cosmological story of **Asar**, **Aset**, **Heru**, and **Sutekh** also know as **Osiris**, **Isis**, **Horus**, and **Set** respectively. The synopsis of the story is:

"Asar is killed by his brother Sutekh, and Asar's body is chopped up into multiple pieces and thrown into the Nile River. Asar's wife Aset searches for her husband's body, and when she finds all of the pieces, she uses Technology to resurrect Osiris and conceive a son by the name of Heru. When Heru grows to adulthood, he confronts Sutekh to avenge the death of his father Asar. During the fight between Heru and Sutekh, Heru loses an eye in the battle, but emerges victorious."

The two principles that emerge from the cosmological story of Asar, Aset, Heru, and Sutekh is that 1) Resurrection can be made possible by way of Technology as in Aset resurrecting her husband Asar, and 2) the product or child born by way of technology (Heru) represents the resurrection of the father (Asar) and is able to accomplish tasks that were not able to be accomplished before. We have seen manifestations of this story play out in movies and media that deal with the themes of Technological Transhumanism. Whether it is in the comics with the **Cyborg Victor Stone**, or in the movies with the **Terminator**, these symbols of Technological Transhumanism are always depicted with **one Human eye** and **one Robot eye** as a reference to the Ancient African deity Heru who was born by way of technology and lost his **eye** in his battle with Sutekh. Moreover, the **right and left eye of Heru** represented **the sun and the moon** respectively, therefore Heru is also symbolic of the **unification of the dualities** which are represented by the sun and the moon or **life and death** in Ancient African cosmology. Since Heru lost an eye in his battle, he could also be considered a **Cyclops** or **"one eyed being"** and then there is another movie character called **RoboCop** who died and was **resurrected by way of technology** and depicted with a helmet which had a **"single eye"** like a **Cyclops** (symbolic of an **"open third eye"**). So we see the same story or the same Program of Heru and Technological Transhumanism running itself on different operating systems and manifesting itself in different forms throughout the ages. Also, the concept of **"Eternal Life"** is an Ancient African Egyptian principle which is symbolized by the **Ankh.**

The book entitled **"P.T.A.H. Technology"** by **African Creation Energy** discusses how the **Ankh** is related to the **flow of electrical current**. Therefore the application of the Ancient African science manifests as Technological Transhumanism and "Eternal Life" made possible by

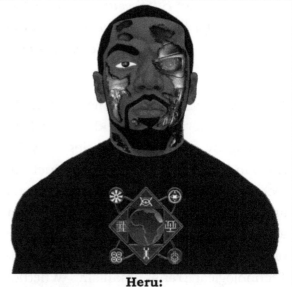

Heru:
Eternal Life (Ankh) by way of Technology,
A "Left Eye" lost in a Battle,
Unification of dual principles

way of a continuous flow of energy and electrical current.

Scientists have still not come to a conclusion about what caused the various genetic leaps from one genus Homo to another. Perhaps Technology is the **missing link** to explain the genetic leaps in the past, in which case Technological Transhumanism has already occurred.

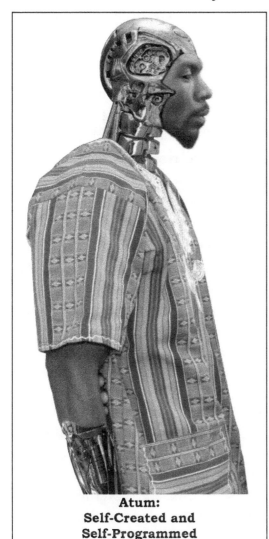

**Atum:
Self-Created and
Self-Programmed**

If one of the characteristics of being Human is to be mortal and die, and one of the characteristics of God is to be Immortal and never die, then Technological Transhumanism is a way to realize one of the characteristics of God and become **God-like**. Sustained life and resurrection after death are realistic possibilities through computer science, cybernetics, and robotics which transcend the frailties and limitations of the Human body. For African people, these sciences are practical applications of traditional African cosmologies and philosophies, and therefore we should not just be the users of these technologies which are created by other groups of people, but rather we should be the creators and programmers of these technologies for the survival, well-being, and improved quality of life for African people as a whole. As the Creators and Programmers of our own African

Transhumanism Technologies, we will also manifesting another Ancient African principle of being "self-created" (**Autopoiesis**) as in the story of the Ancient Egyptian deity **Atum** (son of Ptah). The African cosmological concept of being "self-created" was also portrayed in the animated film entitled "**Kirikou and the Sorceress**" when Kirikou metaphorically "spoke from the womb and brought himself into the world".

"**Digital Immortality**" may be possible sooner than Immortality is made possible by way of Technology in the form of Robotics and Cybernetics. The idea behind Digital Immortality is that all of a person's thoughts, memories, predictable responses, and thought patterns can be uploaded into a computer via a process called "**Mind Uploading**" where the life experiences of the individual would "live forever" creating a "virtual immortal person". An example of Digital Immortality was depicted in the year 2011 movie entitled "**Source Code**" where a soldier killed in war was kept alive, seemingly forever, by having his memories and "sense of self" placed inside of a computer. Digital Immortality can be likened to the African concept of "Spiritual Possession" where a person's "spirit and soul" could possess a computer and dwell in the computer.

As science progresses and Humanity begins to realize and actualize the ideas of Transhumanism in the coming years, all old and non-practical philosophies, ideologies, religions, and socioeconomic systems will be either destroyed or reorganized. The need for a "science-based" religion and a "science-based" government will arise. The Natural Resources needed by Transhumans for survival will differ greatly from the Natural Resources needed for modern Human survival. Modern Humans require the consumption of food for energy, whereas Transhumans could potentially utilize solar energy and would no longer need to consume physical food. Potentially, Transhuman bodies will not need to breathe air, which will free or Liberate people to live on other planets and travel to parts of the Universe where modern Humans could not currently reside. Economic systems will have to be restructured with the possibility of the need for money eliminated. A "science based" **Technocracy** government may be established with scientist, mathematicians, and engineers as world leaders and policy makers rather than lawyers and businessmen. With death eliminated and Transhumans living in outer space, religion will take on a new form with Science as the new religion, Scientist as "Prophets", and Scientific Theories as "Theos" or "God". Transhumanism could even be incorporated into the **rituals of secret societies**. There are countless organizations that use **"death and rebirth"** as a metaphor representing the end of a person's old self identity and the rebirth of a new identity for the person in the organization. However, as Transhumanism ideas are realized, the metaphorical "death and rebirth" rituals may be replaced with rituals where people will be actually killed as a Human and reborn and resurrected as a Cybernetic Transhuman. In this Transhuman state, the individual may be immortal, and thus they are reborn and resurrected as a "God"; this paradigm will give new meaning to the phrase "*What Fools these Mortals be*". With the elimination of death, the elimination of the need for food, and the elimination of the need for money, crime will also be eliminated. There are many possible **positive Utopian** outcomes to the realization of Transhumanism ideas, and since Transhumanism ideas originated in African philosophies and cosmologies, Transhumanism could mean a resurrection and rebirth of traditional African customs if viewed and utilized properly. However, there are also many **negative Dystopian** outcomes which could arise from Transhumanism ideas. It is important for African people to be creators and developers at the forefront of any Transhumanism science and technology so that we can influence and guide Transhumanism to be positive and benefit African people as a whole.

1001 – Rise of the NUBOTS:
An Afro-futuristic Story

W W W . A F R I C A N C R E A T I O N E N E R G Y . C O M

R I S E   O F   T H E
NUBOTS
NUBIAN ROBOTS OF SOUND RIGHT REASONING

The year is **2043** A.D. For decades Africa has been the technological waste dumping ground of the world. It is common place to witness old discarded laptops, hard drives, motherboards, monitors, cell phones, automobiles, televisions, batteries, microwaves, and other unwanted electronic and mechanical waste from Europe, Asia, and the Americas lining the city streets of almost every major city in every African country. The African people have long wanted the pollution and exploitation of their land to end. However, the corrupt governments that were put in place as post-colonial **"Avatars"** of the old European colonizers have ignored the demands of the people in favor of the foreign interest. Unbeknownst to the rest of the world, in the city of **Gaborone**, in the country of **Botswana**, a young inventor by the name of **Sekou Kuumba** [pronounced SAY-KHUU KHUU-OOM-BAA] is on the verge of creating a technology that will not only improve the quality of life for African people, but will ultimately transform the social, political, and economic paradigm of the Earth and the entire Universe.

There is an old saying that states "one man's trash is another man's treasure" and for Sekou Kuumba and many of the people throughout Africa, this statement is very true. For generations Africans have utilized their creativity and ingenuity to make use of the technological waste that has been flooding their continent. Sekou Kuumba has developed solar panel fields to provide his village with free electricity and power, he has transformed old refrigerators and air conditioners into devices which convert air into flowing running water for his village, and he has built and programmed miniature robots that can perform agricultural tasks such as planting and harvesting crops to eat, as well as perform building tasks such as constructing places for members of his tribe to live. With all that Sekou Kuumba has developed to help to improve the quality of life of his people, he has always been troubled with the reality that with all he has done to make life better for his people, that one day him and his people will die. Death and Liberation are the only problems that plague his people which Sekou Kuumba has been unable to create a solution for...until now.

One day while Sekou Kuumba was working in his laboratory, he heard a knock on the door which sounded like something was urgent and in despair. *BOOM! BOOM! BOOM! BOOM! BOOM!* When Sekou went to see who was knocking on the door, Sekou saw that his wife had already answered the door. While Sekou Kuumba has dedicated his life to Science, Sekou's wife, **Ishi Kuumba** [pronounced EE-SHEE KHUU-OOM-BAA], is an African fetish priestess who dedicated her life's work to the study and development of her craft. Sekou's wife Ishi Kuumba has studied every traditional African spiritual system on the planet and is considered a powerful spiritual master and well respected amongst her people. On this particular day, a woman had come to their home seeking the spiritual aid of Sekou's wife. The woman who had been knocking on the door was in tears. The woman explained to Ishi Kuumba that she was being physically and verbally abused at home. The woman wanted Ishi Kuumba to create a fetish for her to stop the abuse. Ishi Kuumba agreed to help the woman, but first asked the woman *"**why did she desire to stay in an abusive relationship?**"* The woman confided in Ishi Kuumba and explained that the reason why she wanted to stay with a man who she knew was abusive to her was because they had children together and because she was so use to having the man provide for her for so long that she was uncertain if she could provide for herself. Ishi Kuumba suggested to the woman that there may be an element of low self-esteem also, and the woman looked down on the floor and the nodded in agreement. Ishi Kuumba created a "fetish doll" for the woman and instructed her that whenever the husband approached her, and she felt threatened like he may abuse her, the woman could prevent the husband from abusing her by controlling the arms of the "fetish doll". Just as the woman received the "fetish doll" from Ishi Kuumba, the woman's Husband came to the home of Sekou and Ishi looking for the woman. The woman's Husband was very angry and irate, yelling and screaming at the woman: *"Where have you* been, I have searched everywhere for you and could not find you." After yelling these words, the woman's Husband raised his hand as if he was going to hit the woman. The woman grabbed the arm of the "fetish doll" given to her by Ishi Kuumba, and made the hand of the fetish doll" slap the fetish doll's own face. Just as the woman's Husband was going to slap the woman, he slapped himself, and knocked himself out. Both Ishi Kuumba and the woman laughed at this incident, and the woman gave Ishi Kuumba her payment, and took her unconscious Husband with her. At times, the scientific work of Sekou Kuumba in robotics was indistinguishable from the spiritual work of Ishi Kuumba in creating fetishes.

Sekou overheard the exchange between the woman and his wife and spoke to his wife Ishi after the woman left. Sekou told Ishi, "*I heard the woman explain to you why she desired to stay in an abusive relationship, and upon hearing her explanation, I finally understood that the reason why our people have remained in an unhealthy relationship and been mistreated by the other Nations of the world*". Ishi Kuumba responded to Sekou, "*Indeed, emotion can cloud reason, because the reasonable course of action would be to leave.*" After hearing these words from his wife, Sekou hugged and kissed her and departed saying "You have just given me an idea."

Sekou returned to his laboratory to work on his new creation which he saw as the solution to liberating his people and solving the many problems they encountered on Earth. Sekou's idea was to create a Robot which only utilized "Sound Right Reason" which could inhabit other planets and moons in the solar system and in the universe and make a way for his people so that the reasonable amongst them could leave their current abusive situation and start over.

Sekou contemplated on what he would call his new creation. He comprehended that there was a relationship between the word "**Robot**" which came from a word meaning "**slave**" and the Hebrew word "**Rabbo**" which meant "**Master**". Sekou Kuumba knew that in Ancient Egypt, the word "**Neb**" meant "**master**" and the word "**Nebo**" and "**Nabu**" meant "**prophet**" and "**wisdom**" respectively. Sekou Kuumba choose the name "**Nubot**" for his new creation because Nubot is an abbreviation for the nano-robots called "nucleic acid robot", but also the word "Nubot" was phonetically similar to the "**Nuba**" and "**Noba**" tribes in Africa, and had ties to the words **Neb** and **Nabu**. But, ultimately, "**Nubot**" would be a short form of "**Nubian Robot**".

Sekou would create the Nubot so that it was covered with **Nano photoelectric cells** which act as **miniature solar panels** that cover the surface area of the body giving it a **Black** appearance and it would absorb **electromagnetic radiation** of all frequencies to continuously acquire energy. Thus, even when the Nubot would be in Darkness, it would still be absorbing the electromagnetic radiation that is present and constantly being charged. The process of converting electromagnetic radiation into usable energy that was developed for the first Nubot created by Sekou Kuumba was

Nubot-1

much like the **photosynthesis** process in plants, and therefore parts of the first Nubot Sekou created **glowed green** like the leaves in plants. Sekou Kuumba created the first Nubot **in his image and after his likeness** to utilize **sound logic** and **right reason** and have dominion in parts of the universe where he could not currently travel. Sekou Kuumba built the Nubot from recycled technological waste, and from the elements of Nature, and inserted within it the spark of electricity, and **Nubot-1** became **activated**.

And the day came where Sekou Kuumba wanted to show his newly created Nubot to the rest of the scientific world. A press conference was held Gaborone, Botswana where thousands of scientists and engineers from around the world came to witness the new creation. The announcer spoke, *"Greetings colleagues. I am pleased to introduce to you today Mr. Sekou Kuumba, a scientist, engineer, and inventor from Botswana who has transformed the technological waste in the city into a fully functioning, autonomous, sentient android."* After the announcer spoke, Sekou walks out with his Nubot-1 to a round of applause from seemingly everyone in the audience. As Nubot-1 walks around the stage and interacts with the crowd, people in the audience give a standing ovation – the delegation from the Americas stands and applauds, the delegation from China stands and applauds, the delegation from India stands and applauds. Everyone in the audience was standing and applauding...everyone that is except for one scientist from the United Kingdom named **Dr. Suoil Leber**. Amidst all of the applause, Dr. Suoil Leber remained seated with his arms crossed as his face turned red until he could not take it anymore, then he stood up, and in his British accent he yelled out *"ENOUGH!"* After Dr. Suoil Leber's outburst, the audience became silent. Dr. Suoil Leber began to speak saying "Why should we pay homage and applaud this mockery of science. He has done nothing but recycle trash, but I have made a far superior creation with my **AMAM-droids**." Just months prior to the unveiling of Sekou Kuumba's Nubot-1, Dr. Suoil Leber presented a robot to the scientific community which he called the "AMAM-droid" which stood for **"Artificial Mechanical Autonomous Movement Android"**. Since it was obvious that Sekou Kuumba's Nubot-1 would be in competition with Dr. Suoil Leber's AMAM-droid, Dr. Suoil Leber felt the need to make disparaging remarks by saying that Sekou's Nubot-1 was a *"walking pile of trash and lacked real ingenuity"*. Dr. Suoil Leber then walked out of the conference, and 200 other scientists who agreed with Dr. Suoil Leber points left with him.

After the fiasco at the conference, Sekou and his Nubot-1 returned to his laboratory, and Sekou was on a new mission to make the next version of the Nubot even better. Sekou had several prototypes in his laboratory from the first iteration of the Nubot creation. Sekou has also been experimenting with using **"Dark Energy"** as a power source. To improve the next version of the Nubot, Sekou Kuumba would incorporate the use of Dark Energy. The Dark Energy infused in the new Nubot caused part of it to **glow** in colors like **indigo** and **violet** like an **ultraviolet "black light"**. Sekou remained diligent at work in his laboratory and no one had seen him outside for weeks which led people to believe that Sekou was up to something groundbreaking. News of Sekou's possible breakthrough with Dark Energy eventually reached all the way to Dr. Suoil Leber in London. When Dr. Suoil Leber heard of the news, he took an army of his AMAM-droids with him to Botswana to try to stop Sekou from succeeding.

Late one night, Sekou was working in his laboratory, still in pursuit of utilizing Dark Energy in his next version of the Nubot. The Nubot was fully formed, and the Dark Energy had been infused throughout the Nubot's structure, and Sekou was just making a few finishing touches when he heard a loud crash through the upstairs window. Sekou's wife came running downstairs into his laboratory and bolting the door shut. Sekou asked Ishi *"what is going on?"* Ishi explained to Sekou that there was an army of Robots being directed by Dr. Suoil Leber that was converging on their house with the intent of killing Sekou. Ishi told Sekou to get on the table next to the "Dark Energy" Nubot and she would chant over him a spell of protection. As the army of AMAM-droids demolished the upstairs of the Kuumba home, Ishi was busy chanting her spell of protection. While Ishi was chanting, Sekou went temporarily unconscious. Eventually one of the AMAM-droids kicked in the door to Sekou's downstairs laboratory. Ishi conceals the "Dark Energy" Nubot with a spell. Dr. Suoil Leber followed behind his AMAM-droids and then questioned Ishi Kuumba saying, *"Where this new invention that I hear Sekou is working on?"* Ishi replied, *"I know nothing of which you speak."* Dr. Suoil Leber said, *"You LIE! No worries, I will just take his Nubot-1."* Dr. Suoil Leber instructed his AMAM-droids to seize the Nubot-1 which was powered off and standing in the corner of the laboratory. Dr. Suoil Leber then said, *"And for you insolence, you will know great pain."* Dr. Suoil Leber then instructed one of his AMAM-droids to kill Sekou. The AMAM-droid approaches the unconscious body of Sekou lying on the table and then shoots a LASER through Sekou's left eye which burns a hole through his head. Ishi screamed and then Dr. Suoil Leber says, *"Now Sekou is no more."* Dr. Suoil Leber gave orders to one of his AMAM-droids saying *"Make her suffer."* One of the AMAM-droids approaches Ishi and lifts her body into the air, and slammed her onto the ground crushing multiple bones in her body. Dr. Suoil Leber and the AMAM-droid left the Kuumba residence with the stolen Nubot-1 and Ishi on the floor of the laboratory paralyzed with internal bleeding near death.

Moments later, Sekou becomes conscious, however something is very different. Sekou looks down and sees his dead body on the floor. He soon realizes that Ishi's spell was not a spell of protection, but rather she transferred Sekou's life force, spirit, soul, and consciousness into the body of the "Dark Energy Nubot". As Sekou scans the room, he finds Ishi's motionless body on the floor barely alive. Sekou, in his new form as the "Dark Energy Nubot", begins to work to work on a mobile life support system for his injured wife Ishi by using the Nubot components available in his laboratory.

Sekou integrates Ishi's body with a Cybernetic Nubot life support system which stabilizes her condition and strengthens her vitality. With the Nubot Cybernetics infused with Ishi's body, her blood becomes electric, which causes parts of her to **glow red**. When Ishi wakes up, she looks at her husband in his new form, and then looks at herself in her new form. Sekou asks her is she ok, and she says yes. Sekou tells Ishi, *"Thank you*

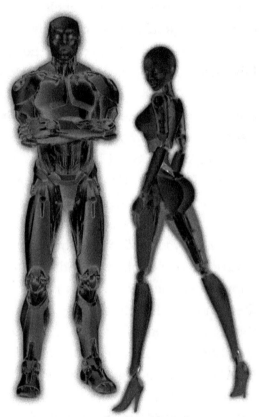

Sekou Nubot and Ishi Nubot in their new form

for saving my life," and Ishi responds to Sekou, *"and thank you for saving my life."* Sekou says, *"Your name Ishi means 'life', and in this form, you will indeed live forever."* Ishi explains to Sekou that the reason why she only concealed the Dark Energy Nubot was because she knew that is where she put Sekou's soul. Ishi regretfully tells Sekou, that the Nubot-1 was stolen during the home invasion. Sekou fears that Dr. Suoil Leber may try to use the Nubot-1 technology with unscrupulous intentions, and therefore decides to embark on a journey to the laboratory of Dr. Suoil Leber to recover the lost Nubot-1.

Meanwhile, Dr. Suoil Leber was busy attempting to **reprogram** the Nubot-1 so that it could be integrated and assimilated into the "**hive-mind**" of his AMAM-droids. Sekou programmed the Nubot-1 with a new programming language that he invented which consisted of both written and verbal commands. The written form of the programming language was based on **binary characters** of the **Nigerian Odus of Ifa**. The verbal form of the programming language was based on the **Xhosa "clicking language"** of the **Khoisan pygmy tribe** in Africa. Therefore, Dr. Suoil Leber could not completely crack the programming code of the Nubot-1 to reprogram it, but could only corrupt the processor of Nubot-1 with a computer virus so that Nubot-1 could not reason soundly and think it was one of the AMAM-droids.

Once the processor of Nubot-1 is corrupted, and Nubot-1 is integrated and assimilated with the other AMAM-droids, all of the robots are kept in a warehouse on the property of Dr. Suoil Leber. The AMAM-droids obtain their power from an electric charging station in the warehouse where each AMAM-droid lines up to get charged. Nubot-1 was designed to be charged from electromagnetic radiation, and even though Nubot-1 can get charged inside of the warehouse, it does not receive the full charge that it could receive if it had exposure to the sun. Also, Nubot-1 is usually last in line to get charged in the warehouse and is generally not maintained in good condition by Dr. Suoil Leber. One day when all of the AMAM-droids and Nubot-1 are lined up in the warehouse for charging and maintenance,

Sekou Nubot and Ishi Nubot break through the walls of the warehouse. Sekou approaches Nubot-1 and instructs Nubot-1 to leave the warehouse and come with him. Nubot-1 responds, "I am not a Nubot, I am an AMAM-droid, and this is where I belong". Sekou realizes that the processor of Nubot-1 has been corrupted and Nubot-1 has lost the ability to reason soundly. Just then, Dr. Suoil Leber entered the warehouse to investigate what was going on. When Dr. Suoil Leber saw Sekou Nubot and Ishi Nubot attempting to take Nubot-1, he ordered the AMAM-droids, and Nubot-1 to *"ATTACK THEM!"*

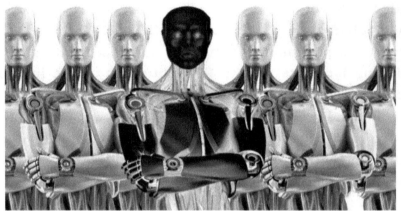

Nubot-1 Integrated into the "Hive Mind" of the AMAM-droids

The AMAM-droid and Nubot-1 began approaching Sekou Nubot and Ishi Nubot in an attack formation. Sekou began shooting "dark energy plasma" blasts from his hands destroying the attacking AMAM-droids. In her Nubot form, Ishi's spiritual powers were amplified, and she began destroying the attacking AMAM-droids seemingly telepathically. When the brawl was over, all of the AMAM-droids were destroyed. Nubot-1 still attempted to attack Sekou Nubot and Ishi Nubot. Ishi took control of the functions of Nubot-1 telepathically causing it to levitate off of the ground. Sekou approached the Nubot-1, and began to correct its programming. When Sekou finished his work, he asked Nubot-1, *"who are you"* and Nubot-1 replied, *"I am that I am, I am Nubot".* Once Sekou knew the correction of Nubot-1 was successful. He turned his attention to Dr. Suoil Leber who was cowering in a corner during this entire ordeal. Sekou said, *"You thought you killed us, be we live on, and are now more powerful. I should kill you but that would be too lenient."* Sekou emitted a "dark energy plasma" blast and proceeded to destroy the laboratory and life's work of Dr. Suoil Leber. Then Sekou and Nubot-1 held Dr. Suoil Leber still, and Ishi transferred the conscious, spirit, and soul of Dr. Suoil Leber into one of the broken AMAM-droid bodies scattered and disheveled on the floor. Then, Sekou Nubot, Ishi Nubot, and Nubot-1 flew off from that place to return to Botswana. Sekou and the Nubots gather some parts from Sekou's old laboratory in Botswana. Using the salvaged parts, the Nubots decide to build a new home on the planet Mars where they will not be under constant attack by humans. However, Sekou does not want to abandon his people. While on Mars, Sekou and the Nubots build a time machine so that they can travel back in time to stop all of the atrocities from happening that have been done to African people over the years.

Powered by Dark Energy,

Utilizing Sound Logic and Right Reason,

the Nubian Cybernetic Android Robots,

called the "Nubots",

travel through time and space,

on a mission of Technological Liberation

from the wicked adverse forces

that are currently plaguing the world,

this is the Rise of the Nubots!

1010 - How to Create an "AfroBot"

The **"AfroBot" Project** was developed for the purpose and intention of providing a project in Robotics and Computer Programming which is grounded in the African Science, African Math, African Engineering, and African Technology information found in the books written by **"African Creation Energy"**. The term "AfroBot" is a combination of the words **"Africa"** and **"Robot"**.

The combination of the words "Africa" and "Robot" in the term "AfroBot" reflects the aim of the **AfroBot Project** to teach Robotics and Computer Programming utilizing symbols, concepts, and constructs from traditional African culture. **Version 1.0** of the AfroBot Project is a design for a simple, **bipedal autonomous obstacle-avoidance walking robot** made out of wood similar to the "fetish" statue robots found in the animated movie entitled *"**Kirikou and the Sorceress**"*. This chapter of the book presents instructions on how to build and program Version 1.0 of the AfroBot Project. Future versions of the AfroBot project will include enhancements such as LASER LEDs for eyes, speech functionality, voice command, and flight capability.

The design of the body of AfroBot Version 1.0 is modified from a project entitled *"Biped walker with 3 servos (Dead Duck Walking)"* by Frits Lyneborg. The modifications to the design include the usage of an **Arduino Microcontroller** and the addition of a **PING))) Ultrasonic sensor**, the addition of arms, and the addition of an **African Baluba mask** for the face/counter-balance weight. The Baluba mask is an African mask of the Luba People, a Bantu tribe native to the Congo region of Central Africa. Also, a new program was written for the Arduino microcontroller to make the robot walk with a new obstacle avoidance routine using the PING))) Ultrasonic sensor.

Materials Needed:

- 6 Paint Sticks (or 560 cm² of Plywood)
- Arduino Uno Microcontroller
- PING))) Ultrasonic sensor
- 3 Servos (either DF15 Metal Gear Servo available at DFRobot.com or Parallax Standard Servo available at Radio Shack or parallax.com)
- 5.5cm x 8.5 cm Bread Board
- Standard jumper wires
- 9 Volt Battery
- 9 Volt Batter Holder
- 10 #6-32x¾" Machine Screws
- 2 #6-32x1" Machine Screws
- 8 #6-32 Machine Screw Nuts
- 10 #2 x 3/8 Wood Screws
- Wood and Plastic Glue
- Thread locking fluid
- Electric Tape

Assembly Instructions:

1. **Leg Panels and Top Stabilizer Panels:** Cut 6 pieces of Plywood which are 3.5cm wide by 6cm long by 0.5cm thick. Drill 2 holes which are 0.5cm in diameter in each piece of wood such that the center of the hole is 1cm from the edge and the two holes are 4cm apart as show in the picture below:

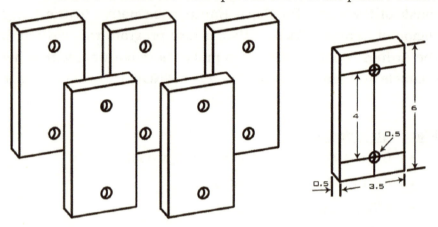

2. **Bottom Stabilizer Panels and Servo Feet Panels:** Cut 4 pieces of Plywood which are 1.5cm wide by 6cm long by 0.5cm thick. Drill 2 holes which are 0.5cm in diameter in each piece of wood such that the center of the hole is 1cm from the edge and the two holes are 4cm apart as show in the picture below:

3. **Feet:** Cut 2 pieces of Plywood which are 6cm wide by 11.5cm long by 0.5cm thick as shown in the picture below:

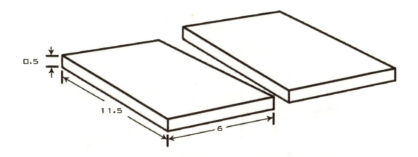

4. **Leg Servo Connectors:** Cut 4 pieces of Plywood which are 2cm wide by 2.5cm long by 0.5cm thick as shown in the picture below

5. Glue each piece of wood cut in step 4 above to the end of a servo as shown in the picture below:

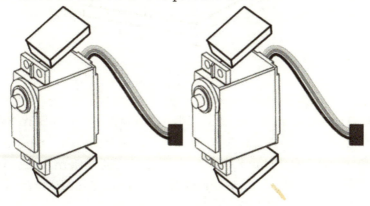

6. **Arms:** Cut two pieces of Plywood which are 3.5cm wide by 14cm long by 0.5cm thick. Remove a 1.5cm wide by 3.5cm long section by cutting the ends of both of these pieces to resemble the picture shown below.

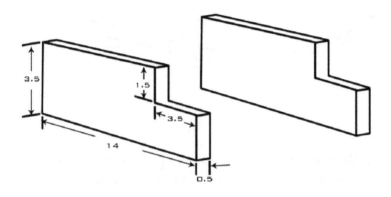

7. **Top Panel to Stabilize and hold Circuitry:** Cut 1 piece of plywood which is 6cm wide by 8cm long by 0.5cm thick as shown in the picture below:

8. **Torso:** Cut 1 piece of Plywood which is 3.5cm wide by 30cm long by 0.5cm thick. The length of this piece of wood could vary. The length of 30cm was chosen because the diameter of the African Baluba mask which is used for the face of the robot is 15cm.

9. **Cut holes for wires in the front leg panels:** In two of the pieces of wood cut in step 1 above, cut holes 1 cm from the drilled hole which are 1.5cm wide by 0.5cm long. These holes will allow a space for the servo wires. These two pieces of wood will be used for the front leg panles.

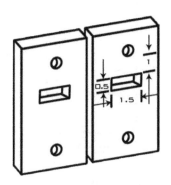

10. **Attach front stabilizer panels to front leg panels:** Use four of the #6-32x¾" Machine Screws and four of the #6-32 Machine Screw Nuts to attach one piece of wood cut in step 1 and one piece of wood cut in step 2 to the two pieces of wood cut in step 9 as shown in the picture below:

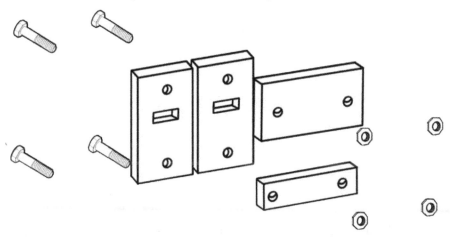

11. **Center all servos at 90 and attach servo horns:** Connect the servos to the Arduino microcontroller using the schematic below:

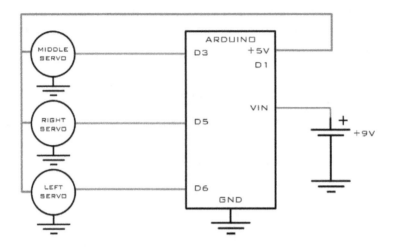

Upload the program below to the Arduino Microcontroller and the 3 servos will be centered to 90 degrees.

```
//=========================================================
//    Arduino Code to set the Servos to 90 degrees
#include <Servo.h>
Servo middleServo;
Servo rightServo;
Servo leftServo;
void setup(){
 middleServo.attach(3);
 rightServo.attach(5);
 leftServo.attach(6);
}
void loop(){
middleServo.write(90);
delay(250);
leftServo.write(90);
delay(250);
rightServo.write(90);
delay(250);
}
```

With the program to center the servos at 90 degrees uploaded to the Arduino Microcontroller, and the 3 servos attached to the Arduino Microcontroller, attach the single-sided servo horns to the two servos with wood glued on their ends from Step 5, and attach the double-sided servo horn to the third servo which does not have wood glued on it as shown in the picture below. The two servos with the wood glued on them which have the single-sided servo horns will serve as the feet servos and the servo with the double-sided servo horn which does not have wood glued on it will serve as the middle servo. If you are using the DF15 Metal Gear Servos, then the dark brown wire will be used for ground, the dark-orange/red wire will be used for +5V, and the light-orange wire will be used to connect to the signal port of the Arduino microcontroller. If you are using Parallax Standard Servos, then the black wire will be used for ground, the red wire will be used for +5V, and the white wire will be used to connect to the signal port of the Arduino microcontroller.

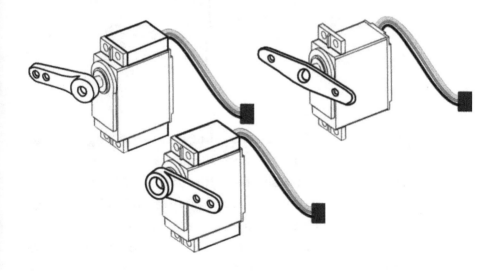

12. Attach middle servo to the rear stabilizers and rear leg panels: Use 2 of the #6-32x1" Machine Screws and 2 of the #6-32 Machine Screw Nuts to attach the middle servo with the double-sided servo horn through the top two drilled holes of 2 of the pieces of wood cut in step 1 positioned vertically with one piece of wood cut in step 1 positioned horizontally as shown in the picture below. Use 2 of the #6-32x¾" Machine Screws along with 2 of the #6-32 Machine Screw Nuts to attach one piece of wood cut in step 2 above through the bottom two drilled holes as shown in the picture below.

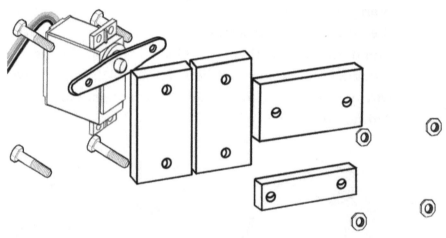

13. Glue the two feet servos between the leg panels assembled in step 10 and step 12: Glue the wood attached to the two servos with the single-sided servo horns to the assemblies constructed in step 10 and step 12 as shown below. The wire end of the feet servos should be glued to the panels with the slots cut for space for the wire and the end of the servo opposite of the wire-end of the feet servos should be glued to the panels which are attached to middle servo as shown in the picture below.

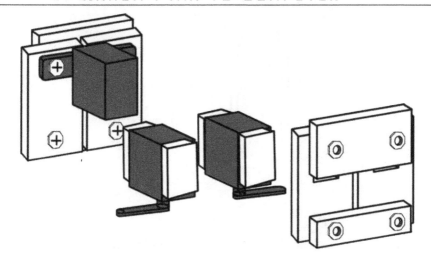

14. **Attach the wood feet panels to the single sided servo horns of the feet servos by glue and/or screw:** Use glue and a #2 x 3/8 Wood Screw to attach one of the pieces of wood cut in step 2 to each of the single sided servo horns of the feet servos as shown in the picture below. There should be between 1cm to 1.5cm of space between the wood feet panels

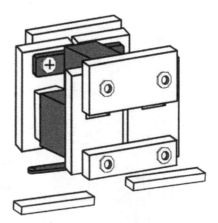

15. **Attach the feet:** Glue the wood feet panels to the pieces of wood which will be used for the feet which were cut in step 3 as shown in the picture below. There should be between 1cm to 1.5cm of space between the feet , and between 3cm to 3.5cm of space between the front end of the feet to the front panels of the legs

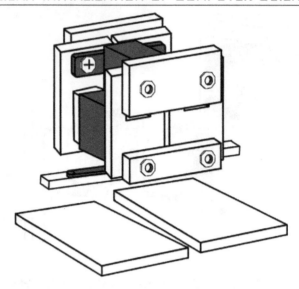

16. Attach the top panel: Glue the top panel piece of wood cut in step 7 to the top of the assembly constructed in step 15. The top panel will be only touching the two pieces of wood which are protruding up in the front and the back as shown in the picture below:

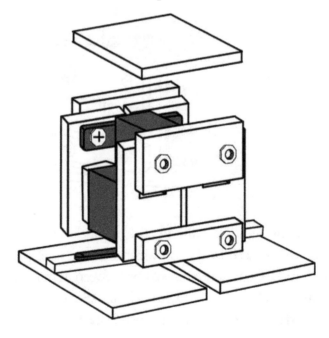

17. Attach the torso by glue and/or screws: Use glue and/or two of the #6-32x¾" Machine Screws to attach the piece of wood which will be used as the torso which was cut in step 8 to the assembly constructed in step 16 as shown in the picture below:

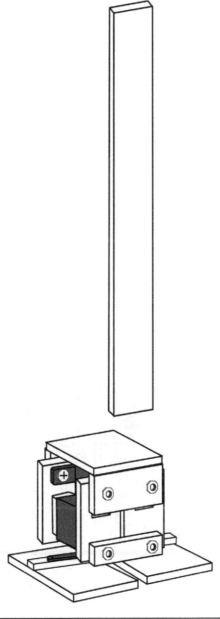

18. Attach the arms: Use glue to attach the 2 pieces of wood which will be used as the arms which were cut in step 6 to the assembly constructed in step 17 as shown in the picture below

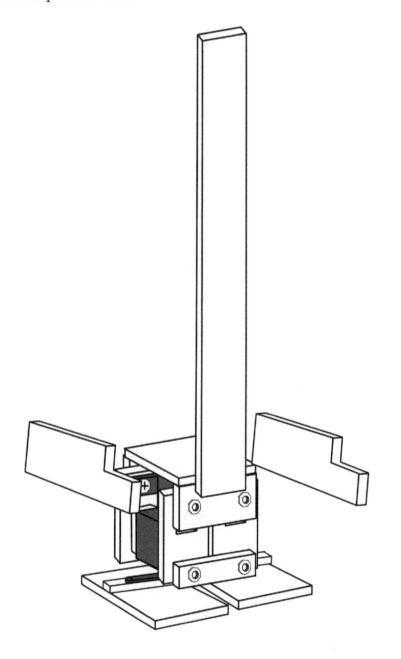

19. Attach the Ping))) Ultrasonic Sensor: Use 2 of the #2 x 3/8 Wood Screws to attach the PING))) Ultrasonic sensor to the torso of the AfroBot as shown in the picture below:

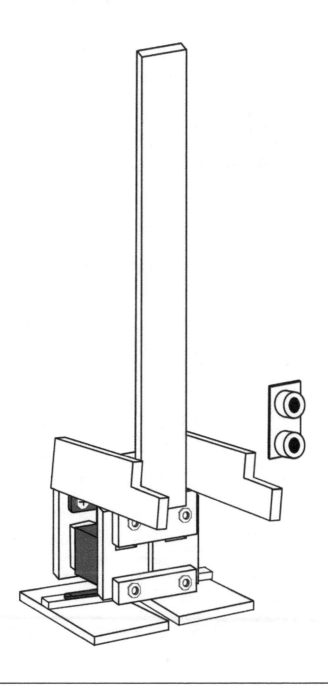

20. Attach the Battery pack: Use electric tape and/or one of the #2 x 3/8 Wood Screws to attach the 9 Volt Batter Holder near the top of the torso of the AfroBot as shown in the picture below:

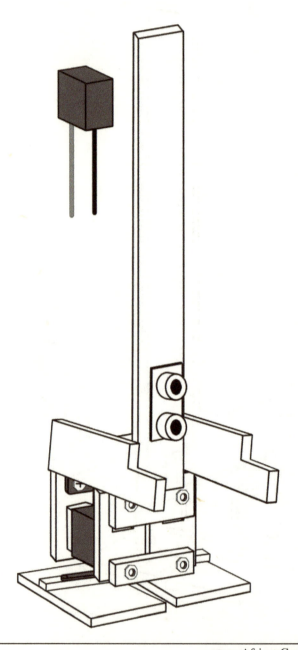

21. Attach the Arduino Microcontroller and breadboard:
Attach the Arduino Microcontroller and breadboard to the top panel piece of wood between the two arms using 4 of the #2 x 3/8 Wood Screws as shown in the picture below

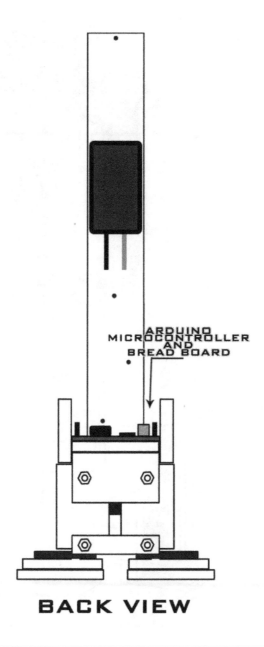

BACK VIEW

22. Attach the African Baluba Mask: Using 2 of the #6-32x¾" Machine Screws, attach the top and bottom of the African Baluba mask to the torso of the AfroBot as shown in the picture below:

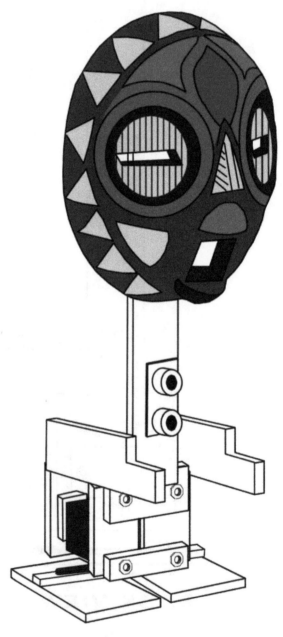

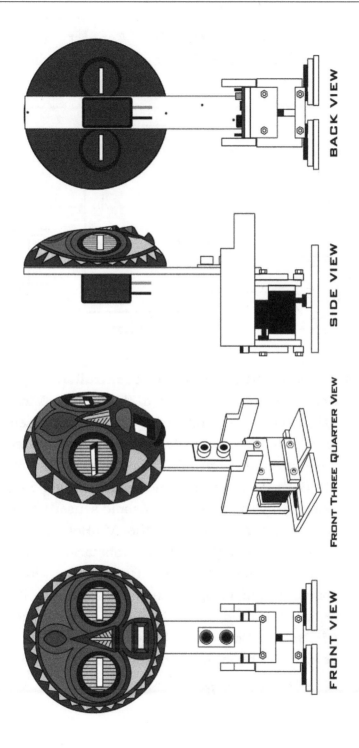

23. Connect the Circuit: Connect the electrical components of the AfroBot using the schematic below:

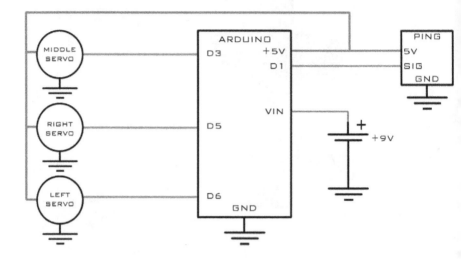

24. Program the Arduino Microcontroller:

Upload the program below to the Arduino Microcontroller using the USB connector. The program enables the AfroBot to autonomously walk and avoid obstacles which are within a 5cm to 20cm distance in front of the AfroBot's PING))) Ultrasonic sensor. When an obstacle is detected by the PING))) Ultrasonic sensor which is within 5cm to 20cm in front of the AfroBot, the AfroBot will begin to turn right until the obstacle is avoided, then continue walking forward. Because the AfroBot detects obstacles and navigates using ultrasound, it is able to walk and avoid obstacles in darkness and in light. **AfroBot is Conscious** because the AfroBot is able to **detect** and be **aware** of obstacles in front of it and avoid the obstacles if they are too close.

```
/*=============================================================
AfroBotWalkItOut.ino  -  AfroBot  Walk  and  Obstacle
Avoidance  Arduino  Code.   Portions  of  this  routine
contain   code   developed   by   Kimmo   Karvinen,   Tero
Karvinen, and Joe Saavedra, 2010
Updated by African Creation Energy 12/12/2012
=============================================================*/
#include <Servo.h>

Servo middleServo;
Servo rightServo;
Servo leftServo;

int walkSpeed = 250; // 0.25 sec
long int duration, distanceInches;
long distanceFront=0; // cm
int startAvoidanceDistance=20; // cm

long microsecondsToInches(long microseconds){
    return microseconds / 74 / 2;
}
long microsecondsToCentimeters(long microseconds){
    return microseconds / 29 /2;
}
long distanceCm(){
    pinMode(pingPin, OUTPUT);
    digitalWrite(pingPin, LOW);
    delayMicroseconds(2);
    digitalWrite(pingPin, HIGH);
    delayMicroseconds(5);
    digitalWrite(pingPin, LOW);

    pinMode(pingPin, INPUT);
    duration = pulseIn(pingPin, HIGH);

    distanceInches = microsecondsToInches(duration);
    return microsecondsToCentimeters(duration);
}
void center(){
    middleServo.write(90);
    delay(walkSpeed);
    leftServo.write(90);
    delay(walkSpeed);
```

```
    rightServo.write(90);
    delay(walkSpeed);
}
void walkForward(){
    // BOTH FEET ON THE GROUND
    middleServo.write(90);
    delay(walkSpeed);
    //SHIFT WEIGHT TO LEFT LEG
    middleServo.write(120);
    delay(walkSpeed);
    rightServo.write(110);
    delay(walkSpeed);
    //TURN LEFT FOOT TO TAKE STEP FORWARD
    leftServo.write(110);
    delay(walkSpeed);
    // BOTH FEET ON THE GROUND
    middleServo.write(90);
    delay(walkSpeed);
    //SHIFT WEIGHT TO RIGHT LEG
    middleServo.write(60);
    delay(walkSpeed);
    leftServo.write(70);
    delay(walkSpeed);
    //TURN RIGHT FOOT TO TAKE STEP FORWARD
    rightServo.write(70);
    delay(walkSpeed);
}
void turnRight(){
    middleServo.write(120);
    delay(walkSpeed);
    leftServo.write(80);
    delay(walkSpeed);
    rightServo.write(90);
    delay(walkSpeed);
    middleServo.write(85);
    delay(walkSpeed);
    rightServo.write(85);
    delay(walkSpeed);
    leftServo.write(90);
    delay(walkSpeed);
}

void setup(){
    // Attach middle servo to Arduino Pin 3
```

```
    middleServo.attach(3);
    // Attach right servo to Arduino Pin 5
    rightServo.attach(5);
    // Attach left servo to Arduino Pin 6
    leftServo.attach(6);
    // PING Ultrasonic sensor to Arduino Pin 1
    pinMode(1, OUTPUT);
    // Center all Servos at Startup
    void center();
}


void loop(){
    distanceFront=distanceCm();
    if (distanceFront > 1){ // Filter 0 error readings
        if (distanceFront <= 5){
            void center(); // Stand Still
            delay(walkSpeed);
        }
        if (distanceFront < startAvoidanceDistance {
            turnRight(); // Turn Right
            delay(walkSpeed);
        } else {
            walkForward(); // walk Forward
            delay(walkSpeed);
        }
    }
}
```

25. **Walk It Out:** After constructing the AfroBot and uploading the walk and obstacle avoidance routine to the Arduino Microcontroller, turn the power on your new creation and observe the first steps of the newly created AfroBot. Use your creativity to make enhancements to the design and programming of AfroBot including a "Walk Backwards" routine, a "Turn Left Routine", and also include additional sensors on the left, right, and back to determine which direction would be the best direction to turn in order to best avoid obstacles.

1011 - About the Author / Programmer

Hello World.

African Creation Energy can scientifically be defined as the Work, Effort, Endeavors, and Activities of African people that cause a movement or change. African Creation Energy is The Energy, Power, and Force that created African people and that African people in turn use to Create. Since African people are the Original people on the planet Earth, it follows from thermodynamics that the Creation Energy of African people is the closest creation Energy of all the people on the Planet to the **Original Creative Energies** that created the Planets, stars, and the Universe. **African Creation Energy** is **Black Power** in the scientific sense of the word "Power", and this book **Radiates** African Creation Energy to be absorbed by the **Black Body**. African Creation Energy has been called by many different names amongst many different groups of African people throughout time. African Creation Energy has been called by the names Ashe, Tumi, Dikenga, Nyama, Nzambi, Amma, Sekhem, NoopooH, and Nuwaupu just to name a few.

The conduit of "African Creation Energy" who has written and authored this book, and other books, goes by the title of **Osiadan Borebore Oboadee** from the Twi language spoken in Ghana West Africa. The Twi word "**Osiadan**" comes from the root words "Si" meaning "Build" and "adan" meaning "Building" with "O-" being a way to denote a "Master". Hence "Osiadan" literally describes a "**Master Builder of Buildings**". Also note the phonetic similarities between the Twi words "Si" and "Adan" and the Ancient Egyptian words "**Sia**" (wisdom) and

"**Aton**" (high noon sun). The Twi word "**Borebore**" comes from the root words "Bo" meaning "Create" and "Re" meaning "to do repetitiously", thus "Borebore" is used to describe a "**Continuous Creation**" or "**Architect**". The word "BoreBore" or "Bore" in Twi is also related to the Hebrew word "**Bara**" meaning "**to begin**" found in the first verse of the first chapter of the Judeo-Christian Bible, and is also related to the Yoruba word "bere" meaning "to begin". Also note the phonetic similarities between the Twi words "Bo" and "Re" and the Ancient Egyptian words "Ba" (soul) and "Re" (sun). The Twi word "**Oboadee**" comes from the root words "Bo" meaning "Create" and "Abode" meaning "Creation" with "O-" being a way to denote a "Master", hence "Oboadee" literally describes a "**Master Creator of Creations**". Oboadee is also pronounced O-Poatee in different African dialects, and is said to derive from the pronunciation of the name of the Ancient African Creation deity PTAH. Osiadan, Borebore, and Oboadee are three principles of **African Creation Energy**. Osiadan Borebore Oboadee is African by blood and lineage; a descendant of the **Balanta-Bassa** and **Djola-Ajamatu** tribes in present day **Guinea-Bissau (Ghana-Bassa)** West Africa. Both the Balanta and Djola tribes migrated to West Africa in Ancient times from the area which is present day **Egypt**, **Sudan**, and **Ethiopia**. Osiadan Borebore Oboadee is a descendant of the Ancient **Napatan**, **Merotic**, **Kushite** Pyramid Builders, and is a Scientist, Engineer, Mathematician, Problem Solver, Analyst, Synthesizer, Artist, Craftsman, and Technologist by education, profession, and Nature. Osiadan Borebore Oboadee has obtained Bachelors and Masters Degrees in the areas of Electrical Engineering, Physics, and Mathematics between the years of 2003 and 2006. Born in the African Diaspora, Osiadan Borebore Oboadee made his first trip to the African continent in the year 2008. Between the years of 2009 and 2010, Osiadan Borebore Oboadee set out to develop, engineer, invent, formulate, build, construct, and create several Technologies (Applications of Knowledge) for the well being of African people worldwide and attempted to radiate the energy that motivated and inspired the development of those technologies in a three part introductory educational series which collectively was entitled "The African Liberation Science, Math, and Technology Project" **(The African Liberation S.M.A.T. Project).** The three books that are part of African Creation Energy's "African Liberation S.M.A.T. project" are:

1. **SCIENCE:** (Knowledge/Information)
   The SCIENCE of Sciences, and The SCIENCE in Sciences

2. **MATHEMATICS:** (Understanding/Comprehension)
   9^{9^9} Supreme Mathematic African Ma'at Magic

3. **TECHNOLOGY:** (Wisdom/Application)
   P.T.A.H. Technology: Engineering Applications of African Science

Osiadan Borebore Oboadee's primary purpose for writing the books of the "African Liberation S.M.A.T. Project" was to motivate the Creative Energies, Minds, and Bodies of African people to go from an inert state of Theory and Speculation to an Active creative state of Development, Creation, and Productivity for the survival and well-being of African people everywhere. It is the goal of African Creation Energy's "African Liberation S.M.A.T. project" to free the minds, energies, and bodies of African people from mental captivity and physical reliance and dependence on inventions and technologies that were not developed or created by, of, and for African people. In 2011, at the age of 30, after writing the books of the "African Liberation S.M.A.T. Project", Osiadan Borebore Oboadee found it necessary to provide evidence of the African Creation Energy Philosophy in Action and Application by building structures and thus embarked upon the project of building a Pyramid and authoring a text entitled **"ARCH I TET: How to Build A Pyramid"** as part of his **30 year "Djed Festival"** of renewal for all eyes to see. In the summer of 2012 Osiadan Borebore Oboadee inscribed a book entitled **"9 E.T.H.E.R. R.E. Engineering"** which was dedicated to having readers better comprehend what they call their "Spirit and Soul" by studying various aspects of Energy including Electricity, Thermodynamics, Hydrodynamics, Electromagnetic Radiation, and Resonant Energy and learning operative use and practical applications of these various types of Energy as a form of "Spirituality".

In Ancient times, the ability to receive revelations in the form of books and convey a message people deemed as "Spiritual" and the ability to see into the future was associated with **Prophets**. In modern times, **Professors** at Universities use statistical models to **"see into the future"** and **make predictions**. Since the word **"Prophet"** and **"Professor"** share a similar etymological meaning and Prophets and Professors both "see into the future", the conduit of "African Creation Energy" who goes by the title of Osiadan Borebore Oboadee, has also adopted the title of **"Prophessor A.C.E."** with the word "Prophessor" being a synthesis of the words "Prophet" and "Professor" and the letters A.C.E. standing for "African Creation Energy". **Prophessor A.C.E.** has worked as a **computer programmer** and **software developer** since 2003 and is fluent in **9 different computer programming languages** including C / C++, HTML, MATLAB, Java, SAS, php, VBA/Visual Basic, FORTRAN, and SQL. Being a **"Futurologist"** or **"Afro-Futurologist"** who uses mathematical models to make predictions and forecasts into the future about

Africans and people of African descent, Prophessor A.C.E. saw that it was necessary to teach the African origin of Computer Science as a way to provide motivation and inspiration for Africans and people of African descent to take part in the creation and development of software, computer programming, and computer-based technologies in the future; thus, this book entitled "**Khnum-Ptah to Computer**" was written to fulfill that objective.

If a Technology, Computer Program, or ANY creation or invention is being used by a group of people, and that Group of People is dependent on that technology for Survival and Well Being, but that Group of people is not in control of the Creation or Production of that Technology, then that Group of People are Literally **SLAVES** to the Creators and the Producers of the Technology. As Africans and people of African descent look to the Future and we see the Rapid Advances of Technology, but we do not see ourselves as the Developers, Creators, Inventors, Designers, or Producers of the Technology, then a sense of Hopelessness, despair, desperation, anger, helplessness, Powerlessness, and feelings of oppression arise. Prophessor A.C.E. has traveled through time with **Afro-Futuristic Consciousness** to show that we as African people have always been **Creators**, **Fashioners**, and **Makers** of Technology and we do indeed have a place as the Developers, Programmers, and Designers of the Programs, Robotics, Artificial Intelligence, and Computers that will shape the Highly Technically Advanced Future. Much like the robot in the Terminator 2 movie was feared and disliked when he traveled back in time to Liberate and aid the people before he was accepted, this new information about the African Initialization of Computer Science and Prophessor A.C.E. will probably be feared and disliked before it is accepted and appreciated.

In the near Future, through Computer technology, the internet, and Smart Phones, the ability to be **conscious**, or **aware**, or **knowledgeable** about something will become trivial, and thus the next level of consciousness, will become **ACTIVENESS** and having the ability to use and put all of the knowledge and information which one is conscious and aware of, into practical application. No longer will "Knowing" be a big deal because almost everyone will "KNOW", the next level will be "**how can what is known be used and applied**". Following the plethora of information presented by the many great African Scholars (who have affectionately been labeled "**MASTER TEACHERS**") who have came to improve the conditions of African people, it is the goal of **Osiadan BoreBore Oboadee** and **African Creation Energy** to be the catalyst in the synthesis, unification, and practical application of the information presented by the great Master Teachers. Thus, it is the aspiration of Osiadan Borebore Oboadee and "African Creation Energy" to be and breed "**Master Technicians**" who TEaCH through Action and Application.

For Africans and people of African Descent, Computer Programming, Computer Science, Binary Code, Artificial Intelligence, Cybernetics, and Robotics are all Concepts and Crafts that are in tune with your Traditional African Culture, Beliefs, Cosmologies, Cosmogonies, and way of life. Get back in tune with your African Self and then take part in the Operative expression of your African Culture by becoming creators and inventors of Computer Programming, Computer Software, Computer Hardware, and Robotics for the overall improvement in the Quality of Life, Survival, and Well Being of African people into the Future.

"Prophessor A.C.E."
(African Creation Energy), also known as
"Osiadan Borebore Oboadee", in year 2011 next to the
Shabaka Stone, the artifact from Ancient Nubia/Kush on
which the Memphite Theology of Ptah is inscribed

www.AfricanCreationEnergy.com

African Creation Energy Books

KHNUM-PTAH To COMPUTER:
The African Initialization of Computer Science
Release Date: 12-12-12

ARCH I TECT:
How to Build A Pyramid
Release Date: 11-11-11

9 E.T.H.E.R. R.E.
Engineering
Release Date: 06-26-12

The SCIENCE of Sciences and The SCIENCE in Sciences
Release Date: 10-10-10

9^{9^9} Supreme Mathematic, African Ma'at Magic
Release Date: 09-09-09

P.T.A.H. Technology: Engineering Applications of African Sciences
Release Date: 05-04-10

1100 - REFERENCES

1. "9^{9^9} Supreme Mathematics African Ma'at Magic" By African Creation Energy

2. "9 E.T.H.E.R. R.E. Engineering" By African Creation Energy

3. "African Philosophy During The Pharaonic Period: 2780 - 330 BC" By Theophile Obenga

4. "African Project Aims To Innovate in Educational Robotics" By Erico Guizzo May 01, 2012

5. "African Student Builds Humanoid Robot From Old TV Parts" By Yuka Yoneda, 04/14/10

6. "ARCH I TECT: How to Build A Pyramid" By African Creation Energy

7. "Australians implant world's first bionic eye" By Associated Free Press Thursday, Aug 30, 2012

8. "Biped walker with 3 servos (Dead Duck Walking)" by Frits Lyneborg http://letsmakerobots.com/node/29379

9. "Black Genesis" By Robert Bauval and Thomas Brophy

10. "Creation records discovered in Egypt: (studies in The Book of the dead)" By George St. Clair

11. "Egyptian Paganism for Beginners: Bring the Gods and Goddesses of Ancient Egypt into Daily Life" By Jocelyn Almond

12. "Make: Arduino Bots and Gadgets" By Kimmo and Tero Karvinen

13. "Mechanical arithmetic, or The history of the counting machine". Chicago: Washington Institute. (1916) By Dorr E. Felt

14. "Odwirafo" By Kwesi Ra Nehem Ptah Akhan, www.odwirafo.com

15. "P.T.A.H. Technology: Engineering Applications of African Sciences" By African Creation Energy

16. "Speaking machines" The parlour review, Philadelphia 1 (3). 20 January 1838. Retrieved 11 October 2010.

17. "Stolen Legacy" by George G.M. James

18. "The SCIENCE of Sciences and the SCIENCE in Sciences" By African Creation Energy

19. "The Temple In Man: Sacred Architecture and the Perfect Man" by Schwaller de Lubicz

1101 – RESOURCES

- African Robotics Network (AFRON) http://robotics-africa.org
- Black Girls Code http://www.blackgirlscode.com
- "Blacks in Science: Ancient and Modern" By Ivan Van Sertima
- "Black Pioneers in Science and Invention" By Louis Haber
- CODE2040 http://code2040.org
- "Dehumanization" by Ben Ammi
- "Dr. York vs. The Computer" by Dr. Malachi Z. York
- "Is Computer Software the New God" by Sadiki Bakari (www.sadikibakari.com)
- "Satellite Terrorism and Cyborg-Genetics Integration of Man and Machine Tome# 22" By Nysut: Amun-Re Sen Atum-Re
- "The Divinity of Afrikan Spiritual Technology vs The Deification of Synthetic Computation" Sadiki Bakari (www.sadikibakari.com)
- "Transhumanism I, Transhumanism II, and Transhumanism III" by Sadiki Bakari (www.sadikibakari.com)
- Terrestrial Extras Theory http://www.terrestrialextras.com

1110 – PHOTO CREDITS

- Asimo, page 103, (www.diseno-art.com/images/asimo-walk.jpg)
- DARwIn-OP, page 103, (www.romela.org)
- KHR, page 103, (www.robotshop.com/Images/big/en/kondo-khr-2hv-humanoid-robot.jpg)
- Sam Todo with SAM10 Robot, page 107, (www.youtube.com/watch?v=sPIq4LbUODk)
- Optimus Prime and the Great Sphinx of Giza, page 81, (www.comicbookmovie.com/images/users/uploads/9186/optimus_sphinx.jpg)
- Palermo stone, page 82, (xoomer.virgilio.it/francescoraf/hesyra/palermo.jpg)
- Darth Vader Helmet, page 111, (http://1.bp.blogspot.com/-4ACdpEzVKQ4/UD6bzPqd6VI/AAAAAAAASc/iLCUcZrr8SM/s1600/masterreplica_darthvaderhelmet1.jpg)

1111 – INDEX

www.ingramcontent.com/pod-product-compliance
Lightning Source LLC
Chambersburg PA
CBHW051238050326
40689CB00007B/970

9 781300 498919